Microprocessor and Microcontroller Interview Questions

A complete question bank with real-time examples

by
ANITA GEHLOT
RAJESH SINGH
P. RAJA
DUSHYANT KUMAR SINGH
PRAVEEN KUMAR MALIK

FIRST EDITION 2020
Copyright © BPB Publications, India
ISBN: 978-93-89845-112

All Rights Reserved. No part of this publication may be reproduced or distributed in any form or by any means or stored in a database or retrieval system, without the prior written permission of the publisher with the exception to the program listings which may be entered, stored and executed in a computer system, but they can not be reproduced by the means of publication.

LIMITS OF LIABILITY AND DISCLAIMER OF WARRANTY
The information contained in this book is true to correct and the best of author's & publisher's knowledge. The author has made every effort to ensure the accuracy of these publications, but cannot be held responsible for any loss or damage arising from any information in this book.

All trademarks referred to in the book are acknowledged as properties of their respective owners.

Distributors:
BPB PUBLICATIONS
20, Ansari Road, Darya Ganj
New Delhi-110002
Ph: 23254990/23254991

DECCAN AGENCIES
4-3-329, Bank Street,
Hyderabad-500195
Ph: 24756967/24756400

MICRO MEDIA
Shop No. 5, Mahendra Chambers,
150 DN Rd. Next to Capital Cinema,
V.T. (C.S.T.) Station, MUMBAI-400 001
Ph: 22078296/22078297

BPB BOOK CENTRE
376 Old Lajpat Rai Market,
Delhi-110006
Ph: 23861747

Published by Manish Jain for BPB Publications, 20 Ansari Road, Darya Ganj, New Delhi-110002 and Printed by him at Repro India Ltd, Mumbai

About the Author

Dr. Anita Gehlot is associated with Lovely Professional University as an Associate Professor with more than ten years of experience in academics. She has **twenty-Five patents** in her account. She has published more than **fifty research papers** in referred journals and conference. She has been awarded with "*certificate of appreciation*" from University of Petroleum and Energy Studies for exemplary work. She has published fifteen books in the area of Embedded Systems and Internet of Things with reputed publishers like CRC/Taylor & Francis, Narosa, GBS, IRP, NIPA, River Publishers, Bentham Science and RI publication. She is an editor to a special issue published by AISC book series, Springer with title *"Intelligent Communication, Control and Devices-2018"*.

Dr. Rajesh Singh is currently associated with Lovely Professional University as a Professor with more than fifteen years of experience in academics. He has been awarded as gold medalist in M. Tech and honors in his B.E. His area of expertise includes embedded systems, robotics, wireless sensor networks, and Internet of Things. He has organized and conducted several workshops, summer internships, and expert lectures for students as well as faculty. He has been honored as keynote speakers and session chair to international/national conferences, faculty development programs and workshops. He has twenty-seven **patents** in his account. He has published around hundred **research papers** in referred journals/conferences.

P. Raja is currently associated with Lovely Professional University as an Assistant Professor with more than eight years of experience in academics. His area of expertise includes MP LAB IDE, IAR, KEIL, Xilinx ISE, Proteus, MODELSIM, QUARTUS, Multisim, Orcad, and CODE VISION AVR, Assembly Language. He has organized and conducted several workshops, summer internships, and expert lectures for students as well as faculty. He has published around fifteen **research papers** in referred journals/conferences.

Dushyant Kumar Singh is an Assistant Professor and Head of Embedded Systems Domain in Lovely Professional University. He has completed his master from Punjab Engineering College, University of Technology, Chandigarh. He has Industrial experience of 2 years and more than 9 years of Teaching experience. He is engaged in Embedded Systems design and IoT systems design since last more than 5 years. His project *"Wireless Router"* is recognized and certified by BSF, Jalandhar and one another project *"PICASSO 4.0"*, IoT based project, has been the winner of the International competition Delta Cup, China in 2017.

Dr. Praveen Kumar Malik is working as a Professor in Dept. of Electronics and Communication, Lovely Professional University, Punjab India. He has written more than 15 papers in different National and International Journals. His major area of interest includes Antenna Design, wireless Communication, and Embedded systems.

Acknowledgements

We acknowledge the support from the publisher, for encouraging our ideas for writing this book and manage this project efficiently.

We are grateful to the honorable Chancellor (Lovely Professional University) Ashok Mittal, Mrs. Rashmi Mittal (Pro Chancellor, LPU), Dr. Ramesh Kanwar (Vice Chancellor, LPU), and Dr. Lovi Raj Gupta (Executive Dean, LPU) for their support and constant encouragement. In addition, we are thankful to our family, friends, relatives, colleagues, and students for their moral support and blessings.

Dr. Anita Gehlot
Dr. Rajesh Singh
P.Raja
Dushyant Kumar Singh
Dr. Praveen Kumar Malik

Preface

The objective of this book is to provide a platform for the students preparing for the interview. The book comprises of questions on *"Microprocessor & Microcontroller"*, covering microprocessor 8085, 8086, microcontroller AVR, PIC and interfacing of peripheral devices.

Errata

We take immense pride in our work at BPB Publications and follow best practices to ensure the accuracy of our content to provide with an indulging reading experience to our subscribers. Our readers are our mirrors, and we use their inputs to reflect and improve upon human errors if any, occurred during the publishing processes involved. To let us maintain the quality and help us reach out to any readers who might be having difficulties due to any unforeseen errors, please write to us at :

errata@bpbonline.com

Your support, suggestions and feedbacks are highly appreciated by the BPB Publications' Family.

Table of Contents

1. **Number Systems** 1
   Introduction to Number systems 1
      Binary number system 1
      Octal number system 2
      Decimal number system 2
      Hexadecimal (hex) number system 3
   Complement 4
      r's complement 4
      (r-1)'s Complement 4
   Number Conversion 5
      Binary to Decimal conversion 5
      Binary to octal conversion 6
      Binary to Hexadecimal conversion 7
      Decimal to Binary conversion 8
      Decimal to octal conversion 9
      Decimal to Hexadecimal conversion 9
      Octal to Binary conversion 10
      Octal to Decimal conversion 10
      Octal to Hexadecimal conversion 10
      Hexadecimal to decimal conversion 11
      Hexadecimal to octal conversion 11
      Hexadecimal to Binary conversion 12

2. **Digital Circuit** 13
   Logic gates 13
   Adder, Subtractor and Multiplier 17
      Half Adder 17
      Full Adder 18
   Encoder and Decoder 20
      Encoder 20
      Decoder 20

	Multiplexer and Demultiplexer	21
	Multiplexer	21
	Demultiplexer	21
	Latches and Flip flop	22
	Flip flop	22
	Register	23
3.	**Microprocessor 8085**	**25**
	Basic terms	25
	Pin diagram and architecture of 8085 Microprocessor	27
	Pin diagram of 8085	27
	Architecture of 8085 Microprocessor	28
	Flag register	28
	Interrupt	29
	Instruction	30
	Op-code	31
	Addressing modes	32
	Logical instructions	34
	Interfacing	35
	Address Decoding Techniques	38
	Absolute decoding/Full Decoding Linear decoding/Partial Decoding	38
	Assembly language program	39
	Assembler directives of 8085	39
	Delay routine	40
4.	**Peripheral Devices and Interfacing**	**41**
	Programmable peripheral Interface (8255)	41
	Programmable interval timer (8253/8254)	42
	Programmable interrupt controller (8259)	43
	Direct Memory Access controller (8257)	44
	Universal Synchronous Asynchronous Receiver Transmitter (8251)	45

5. **AVR ATmega32** 47
   Introduction 47
       Features of ATmega32 is a 40-pin IC: 47
       Characteristics of ATmega32 ADC: 50

6. **Interfacing of Input/Output Device** 51
   Light Emitting Diode (LED) 51
   Seven Segment 51
   Liquid Crystal Display 52
       Pin diagram of 16x2 LCD 53
       Commands used for LCD initialization are: 53
       Motors 54

7. **Exercise** 55
   Descriptive Type Questions 55

8. **Multiple Choice Questions** 143

Chapter 1
Number Systems

Introduction to Number systems

Number systems are used to reparesent the information in various systems. Electronic and Digital systems may use a variety of different number systems, (Decimal, Hexadecimal, Octal, Binary).

A number **N** in base or radix b can be written as:

$$(N)_b = d_{n-1} d_{n-2} \text{-----} d_1 d_0 . d_{-1} d_{-2} \text{-----} d_{-m}$$

In the above:

- d_{n-1} to d_0 is integer part

follows a radix point

- d_{-1} to d_{-m} is fractional part.

Here:

- d_{n-1} = Most significant bit (MSB)
- d_{-m} = Least significant bit (LSB)

Number Systems					
S. No.	Parameters	Binary	Octal	Decimal	Hexadecimal
1	Base	2	8	10	16
2	Symbol/Digits	0 & 1	0 – 7	0 – 9	0– 9 & A – F

Binary number system

The binary number system is another way to represent quantities. The binary number has only two digits 0 and 1.

Weight structure of a binary system is:

$2^{n-1}\ldots\ldots\ldots 2^3\ 2^2\ 2^1 2^0\ .\ 2^{-1}\ 2^{-2}\ 2^{-3}\ \ldots\ldots\ 2^{-n}$

Where:

- **n:** The number of bits from the binary point
- **.:** Binary point
- **2:** Base of binary number system.

Example: $(11011)_2$

Octal number system

The number system whose **base is 8** is known as the **octal number system**. The **base 8 means** the system uses **eight digits** from 0 to 7. The next digit in octal number is represented by 10, 11, 12, 13, 14, 15, 16, 17 which represents the decimal digits 8, 9, 10, 11, 12, 13, 14, 15.

Weight structure of an octal system is:

$8^{n-1}\ldots\ldots\ldots 8^3\ 8^2\ 8^1 8^0\ .\ 8^{-1}\ 8^{-2}\ 8^{-3}\ \ldots\ldots\ 8^{-n}$

Where:

- **n:** positional of octal number system
- **. :** Octal point
- **8:** Base of octal number system.

Example: $(745)_8$

Decimal number system

The number system whose **base is 10** is known as the **decimal number system**. The **base 10 means** the system uses **ten digits** from 0 to 9.

Weight structure of a decimal system is:

$10^{n-1}\ldots\ldots\ldots 10^3\ 10^2\ 10^1\ 10^0\ .\ 10^{-1}\ 10^{-2}\ 10^{-3}\ \ldots\ldots\ 10^{-n}$

Where:

- **n:** positional of Decimal number system
- **. :** Decimal point
- **10:** Base of Decimal number system.

Example: $(7459)_{10}$

Hexadecimal (hex) number system

The number system whose **base is 16** is known as the **Hexadecimal number system**. The **base 16** means the system uses **ten digits** from 0 to 9, A – F.

Weight structure of a decimal system is:

$16^{n-1} \ldots \ldots 16^3 \; 16^2 \; 16^1 16^0 \; . \; 16^{-1} \; 16^{-2} \; 16^{-3} \ldots \ldots 16^{-n}$

Where:

- **n:** positional of Hexadecimal number system
- **. :** Hexadecimal point
- **16:** Base of Hexadecimal number system.

Example: $(75AC)_{16}$

The *table1.1* shows the decimal, binary, octal, and hexadecimal numbers from 0 to 15 and their equivalent binary number.

Table 1.1: Number system equivalent

Decimal	Binary	Octal	Hexadecimal
0	0000	0	0
1	0001	1	1
2	0010	2	2
3	0011	3	3
4	0100	4	4
5	0101	5	5
6	0110	6	6
7	0111	7	7
8	1000	10	8
9	1001	11	9
10	1010	12	A
11	1011	13	B
12	1100	14	C
13	1101	15	D
14	1110	16	E
15	1111	17	F

Complement

Complements are used in digital computers to simplify subtraction. The complements are classified in two types based on the number systems base.

a. r's complement or Radix complement.
b. (r-1)'s complement or diminished complement.

r's complement

If N is a positive number to base r with integer part of **n** digit, the r's complement of N is defined as $r^n - N$.

r's complement = $r^n - N$

Where:
- **r:** base of the number system
- **n:** number of integer digit
- **N:** given positive number

Or

r's complement = (r-1)'s complement + 1

Question 1.1: Find 2's complement of 11011 binary number

No of integer digits = 5

Base r = 2

$$
\begin{aligned}
\text{2's complement of 11011} &= 2^5 - 11011 \\
&= (32)_{10} - 11011 \\
&= 100000 - 11011 \\
&= 00101
\end{aligned}
$$

(r-1)'s Complement

If N is a positive number in base r with **n** integer digits and **m** fraction digits, **(r-1)'s** complement is defined as:

$$[(r^n - 1) - N] \text{ or } (r^n - r^{-m} - N)$$

Where:
- **r:** base of the given number system
- **n:** number of integer digits
- **N:** given positive number

- **m:** number of fraction digits

Question 1.2: Find 1's complement of 10110 binary number.

Number of integer digit = 5

Base   r = 2

1's complement of 10110   = $[(2^5 - 1) - 10110]$
                                     = $[((32)_{10} - 1) - 101101]$
                                     = (100000 − 1) − 101101
                                     = 01001

Number Conversion

Binary to Decimal conversion

Steps:

1. Write the binary number
2. Write the weights 2^0 2^1 2^2 2^3 etc., under the binary digits starting with the bit on right hand side.
3. Cross out weights under zeros
4. Add the remaining weights.

Question 1.3: Convert the given binary number into decimal number $(10011)_2 = (?)_{10}$

First, multiply all the digits in the number by 2 and add them:

$$(1*2) + (0*2) + (0*2) + (1*2) + (1*2)$$

Give powers to 2 starting from 0 from right to left.

$$(1*2^4) + (0*2^3) + (0*2^2) + (1*2^1) + (1*2^0)$$

That's all the formula part.

Now, convert the powers of 2 to the numbers.

$$(1*16) + (0*8) + (0*4) + (1*2) + (1*1)$$

Summing up all, gives the answer as:

$$16 + 0 + 0 + 2 + 1 = 19$$

So, $(19)_{10}$ is the decimal equivalent of the given binary number $(10011)_2$.

Question 1.4: Convert the given binary number into decimal number $(110111)_2 = (?)_{10}$

$(1*2^5) + (1*2^4) + (0*2^3) + (1*2^2) + (1*2^1) + (1*2^0)$

$= 32 + 16 + 0 + 4 + 2 + 1$

$= 55$

So, $(55)_{10}$ is the decimal equivalent of the given binary number $(110111)_2$.

Question 1.5: Convert the given binary number to decimal number $(1101)_2 = (?)_{10}$

Binary number	1 1 0 1
Write weights	8 4 2 1
Cross eights	8 4 2 1
Add weights	8 + 4 + 1 = 13

Question 1.6: Convert the given binary number to decimal number $(1011.101)_2 = (?)_{10}$

1. Split non-fractional part and fractional part

Non fractional part: $(1011)_2$

Fractional part: $(.101)_2$

2. Binary to decimal conversion for non-fractional part

$(1*2^3) + (0*2^2) + (1*2^1) + (1*2^0)$

$= 8 + 0 + 2 + 1$

$= (11)_{10}$

3. Binary to decimal conversion by putting negative power for fractional part

$(1*2^{-1}) + (0*2^{-2}) + (1*2^{-3})$

$= (1* 1/2) + 0 + (1*1/8)$

$= (1/2) + (1/8) = 5/8 = 0.625$

4. So, we got $(0.625)_{10}$, which is the decimal equivalent of the binary fraction $(.101)_2$. Therefore, the decimal form of the binary number $(1011.101)_2$ is $(11.625)_{10}$.

Binary to octal conversion

Question 1.7: Convert the given binary number to decimal number $(10111101)_2 = (?)_8$

In this the given number is grouped in bits of threes, and then each group is converted to its decimal.

101 → 5

111 → 7

010 → 2

So, **(275)₈** is the octal equivalent of the given binary number **(10111101)₂**.

Question 1.8: Convert the given binary number to decimal number **(1011100011)₂ = (?)₈**

In this the given number is grouped in bits of threes, and then each group is converted to its decimal.

011 → 3

100 → 4

011 → 3

001 → 1

So, **(1343)₈** is the octal equivalent of the given binary number **(1011100011)₂**.

Binary to Hexadecimal conversion

Question 1.9: Convert the given binary number to Hexadecimal number **(1011100011)₂ = (?)₁₆**

In this the given number is grouped to bits of fours, and then each group is converted to its Hexadecimal.

    0011 → 3

    1110 → 14 → E

    0010 → 2

So, **(2E3)₁₆** is the Hexadecimal equivalent of the given binary number **(1011100011)₂**.

Question 1.10: Convert the given binary number to Hexadecimal number **(111001100011101)₂ = (?)₁₆**

In this the given number is grouped to bits of fours, and then each group is converted to its Hexadecimal.

    1101 → 13 → D

    0001 → 1

    0011 → 3

    0111 → 7

So, $(731D)_{16}$ is the Hexadecimal equivalent of the given binary number $(111001100011101)_2$.

Decimal to Binary conversion

Question 1.11: Convert the given decimal number to binary number $(548)_{10} = (?)_2$

In this conversion, we divide the decimal number continuously by 2, note down its remainder and continue with the dividing of number until the quotient get 0.

548/2;	Quotient = 274;	remainder = 0
274/2;	Quotient = 137;	Remainder = 0
137/2;	Quotient = 68;	Remainder = 1
68/2;	Quotient = 34;	Remainder = 0
34/2;	Quotient = 17;	Remainder = 0
17/2;	Quotient = 8;	Remainder = 1
8/2;	Quotient = 4;	Remainder = 0
4/2;	Quotient = 2;	Remainder = 0
2/2;	Quotient = 1;	Remainder = 0
1/2;	Quotient = 0;	Remainder = 1

We have to take (write) down to up as MSB to LSB correspondingly.

So, $(1000100100)_2$ is the Binary equivalent of the given decimal number $(548)_{10}$.

Question 1.12: Convert the given decimal number to binary number $(41.68)_{10} = (?)_2$

For fractional part:

For fractional decimal numbers, multiply it by 2 and record the carry in the integral position.

For integral part is remaining same:

41/2;	Quotient = 2;	Remainder = 1
20/2;	Quotient = 10;	Remainder = 0
10/2;	Quotient = 5;	Remainder = 0
5/2;	Quotient = 2;	Remainder = 1
2/2;	Quotient = 1;	Remainder = 0

1/2;	Quotient = 0;	Remainder = 1

For fractional part:

0.68*2;	= 1.36	Carry of 1
0.36*2;	= 0.72	Carry of 0
0.72*2;	= 1.44	Carry of 1
0.44*2;	= 0.88	Carry of 0
0.88*2;	= 1.76	Carry of 1

For fractional part we need take (Write) from top to bottom.

So, $(101001.10101..)_2$ is the Binary equivalent of the given decimal number $(41.68)_{10}$.

Decimal to octal conversion

Question 1.13: Convert the given decimal number to octal number $(458)_{10} = (?)_8$

In decimal to octal conversion, number is continuously divided by 8 and the remainder is recorded until the quotient is 0.

458/8;	Quotient = 57;	Remainder = 2
57/8;	Quotient = 7;	Remainder = 1
7/8;	Quotient = 0;	Remainder = 7

So, $(712)_8$ is the Octal equivalent of the given decimal number $(458)_{10}$.

Decimal to Hexadecimal conversion

Question 1.14: Convert the given decimal number to Hexadecimal number $(458)_{10} = (?)_{16}$

In decimal to hexadecimal conversion, number is continuously divided by 16 and the remainder is recorded until the quotient is 0.

458/16;	Quotient = 28;	Remainder = 10 (In hex is A)
28/16;	Quotient = 1;	Remainder = 12 (In hex is C)
1/16;	Quotient = 0;	Remainder = 1 (In hex is 1)

So, $(1CA)_{16}$ is the Hexadecimal equivalent of the given decimal number $(458)_{10}$.

Octal to Binary conversion

Question 1.15: Convert the given octal number to binary number $(472)_8 = (?)_2$

In octal to binary conversion, number can be converted to binary by converting each octal digit to its 3-bits binary equivalent.

4 → 100

7 → 111

2 → 010

So, $(100111010)_2$ is the binary equivalent of the given octal number $(472)_8$.

Octal to Decimal conversion

Question 1.16: Convert the given octal number to Decimal number $(472)_8 = (?)_{10}$

An octal number can easily be converted to decimal number by multiplying each octal digit by its positional weight.

$(472)_8 = (4*8^2) + (7*8^1) + (2*8^0)$

$= (256) + (56) + (2)$

$= (314)_{10}$

So, $(472)_8$ is the binary equivalent of the given octal number $(314)_{10}$.

Octal to Hexadecimal conversion

Question 1.17: Convert the given octal number to Hexadecimal number $(472)_8 = (?)_{16}$

An octal number can easily be converted to decimal number by multiplying each octal digit by its positional weight.

$(472)_8 = (4*8^2) + (7*8^1) + (2*8^0)$

$= (256) + (56) + (2)$

$= (314)_{10}$

So, $(472)_8$ is the binary equivalent of the given octal number $(314)_{10}$.

Number System 11

Hexadecimal to decimal conversion

Question 1.18: Convert the given Hexadecimal number into Decimal number $(24.6)_{16} = (?)_{10}$

A hexadecimal can be converted to decimal number by multiplying each decimal number by its hexadecimal positional weight.

$(24.6)_{15} = (2*16^1) + (4*16^0) + (6*16^{-1})$

$= (32) + (4) + (0.375)$

$= (36.375)_{10}$

So, $(36.375)_{10}$ is the Decimal equivalent of the given Hexadecimal number $(24.6)_{16}$.

Question 1.19: Convert the given Hexadecimal number into Decimal number $(107)_{16} = (?)_{10}$

A hexadecimal can be converted to decimal number by multiplying each decimal number with its hexadecimal digit by its positional weight.

$(107)_{16} = (1*16^2) + (0*16^1) + (7*16^0)$

$= (256) + (0) + (7)$

$= (263)_{10}$

So, $(263)_{10}$ is the Decimal equivalent of the given Hexadecimal number $(107)_{16}$

Hexadecimal to octal conversion

Question 1.20: Convert the given Hexadecimal number into Decimal number $(45A8)_{16} = (?)_{8}$

In Hexadecimal to octal conversion, number is converted into binary by converting each octal digit to its 3-bits binary equivalent.

$(45A8)_{16} = (100010110101000)_2$

  000 → 0

  101 → 5

  110 → 6

  010 → 2

  100 → 4

So, $(42650)_8$ is the Decimal equivalent of the given Hexadecimal number $(45A8)_{16}$.

Hexadecimal to Binary conversion

Question 1.21: Convert the given Hexadecimal number into binary number $(45A8)_{16} = (?)_2$

In Hexadecimal to binary conversion, number can be converted to binary by converting each hexadecimal digit to its 4-bits binary equivalent.

$4 \rightarrow 0100$

$5 \rightarrow 0101$

$A \rightarrow 1010$

$8 \rightarrow 1000$

So, $(100010110101000)_2$ is the binary equivalent of the given octal number $(45A8)_{16}$.

Chapter 2
Digital Circuit

Logic gates

Not gate

A *'NOT'* gate is having one input and one output, output is the complement of input. A **NOT** gate is also called as **inverter**.

Symbol:

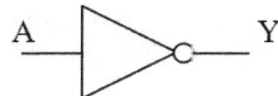

Truth Table:

Input (A)	Output (A)
0	1
1	0

OR Gate

'Or' gate is having two or more than two inputs and one output. The output of OR gate is low when both inputs are low otherwise high.

Symbol:

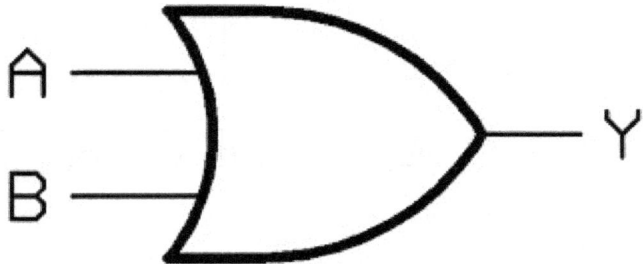

Truth Table:

Input		Output
A	B	Y
0	0	0
0	1	1
1	0	1
1	1	1

AND Gate

'*AND*' gate is having two or more than two inputs and one output. The output of **AND** gate is high when both the inputs are high otherwise low.

Symbol:

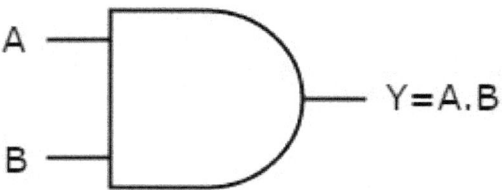

Truth Table:

Input		Output
A	B	Y
0	0	0
0	1	0
1	0	0
1	1	1

NAND Gate

'*NAND*' gate is having two or more inputs and one output. The output of **NAND** gate is low when both the inputs are high otherwise the output will be high.

Symbol:

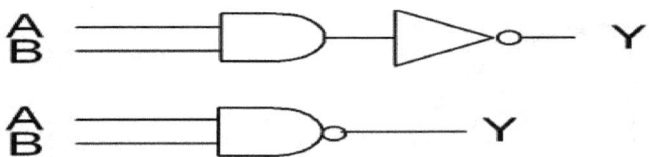

Truth Table:

Input		Output
A	B	Y
0	0	1
0	1	1
1	0	1
1	1	0

NOR Gate

'*NOR*' gate is having two or more than two inputs and one output. The output of **NOR** gate is high when both the inputs are low otherwise the output will be high.

Symbol:

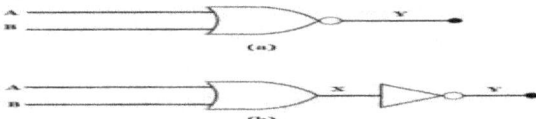

Truth Table:

Input		Output
A	B	Y
0	0	1
0	1	0
1	0	0
1	1	0

ExOR Gate

'ExOR' gate is having two or more than two inputs and one output. The output of **ExOR** gate is high when both the inputs are different otherwise the output will be low.

Symbol:

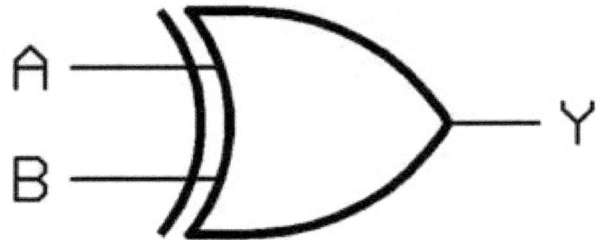

Truth Table:

Input		Output
A	B	Y
0	0	0
0	1	1
1	0	1
1	1	0

ExNOR Gate

'*ExNOR*' gate is having two or more than two inputs and one output. The output of **ExNOR** gate is high when both the inputs are same otherwise it will be low.

Symbol:

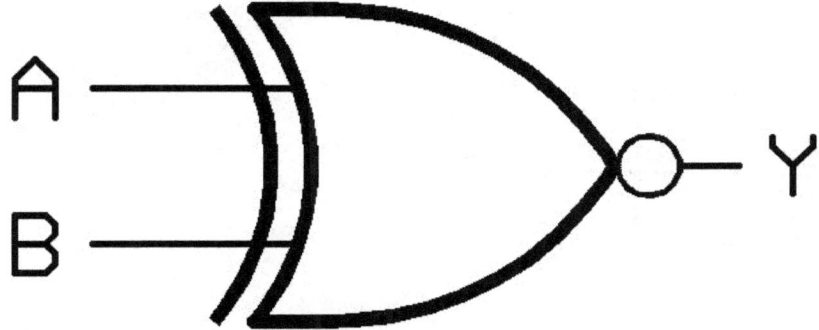

Truth Table:

Input		Output
A	B	Y
0	0	1
0	1	0
1	0	0
1	1	1

Adder, Subtractor and Multiplier

Half Adder

Half adder is a **combinational circuit**, which is used for performing the logical addition on two-bit numbers. A half adder performs the addition of two inputs, and it produces two outputs namely **sum** and **carry**. In the truth table, **A** and **B** is termed as inputs while the outputs **sum** and **carry** are named as **S** and **C** respectively.

Block diagram:

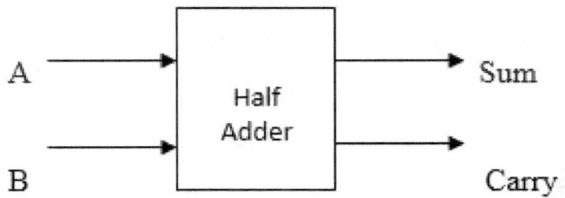

Circuit diagram:

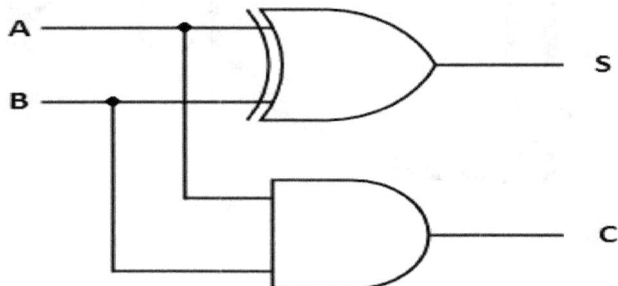

Truth table:

Input		Output	
A	B	Sum	Carry
0	0	0	0
0	1	1	0
1	0	1	0
1	1	1	1

Boolean expression:

Sum = A'B + AB'

Carry = AB

Full Adder

Full adder performs addition operation when the augend and addend number contain more than 2 digits. The **carry** obtained from the addition of 2 bits is added to the next higher pair of significant bits. Here, the addition operation involves 3 bits, which are

augend bit, **addend** bit and **carry** bit respectively.

The outputs of the full adder are also referred as **sum** and **carry**.

Block diagram:

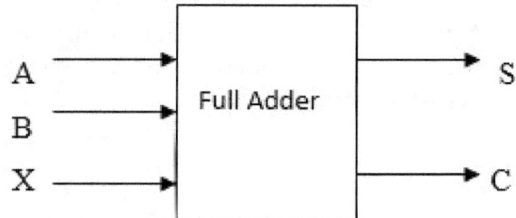

Circuit diagram:

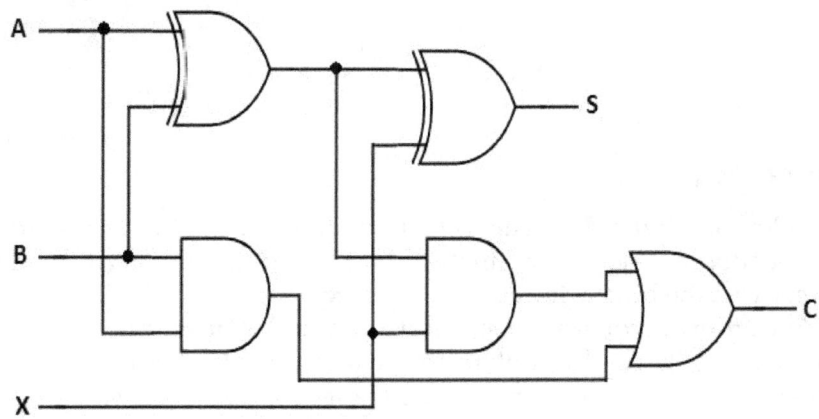

Truth Table:

Input			Output	
A	B	X	S	C
0	0	0	0	0
0	0	1	1	0
0	1	0	1	0
0	1	1	0	1
1	0	0	1	0
1	0	1	0	1
1	1	0	0	1
1	1	1	1	1

Encoder and Decoder

Encoder

An encoder is a **combinational circuit** that converts binary information in the form of a **2N** input lines into **N** output lines, which represent N bit code for the input. For simple encoders, it is assumed that only one input line is active at a time.

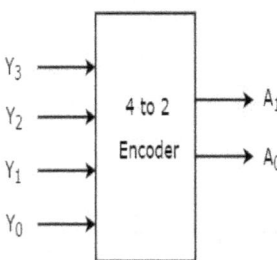

Decoder

Decoders are **digital ICs** which are used for decoding. In other words, the decoders **decrypt** or obtain the actual data from the received code, i.e. convert the binary input at its input to a form, which is reflected at its output. It consists of **n** input lines and **2^n** output lines. A decoder can be used to obtain the required data from the code or can also be used for obtaining the parallel data from the serial data received.

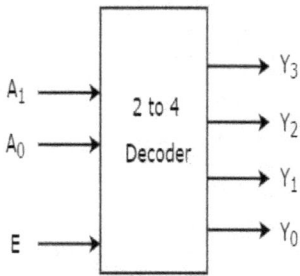

Multiplexer and Demultiplexer

Multiplexer

Multiplexer means many into one. A multiplexer is a circuit used to select and route any one of the several input signals to a signal output. A simple example of a non-electronic circuit of a multiplexer is a single pole multi position switch.

Multi-position switches are widely used in many electronics circuits. However, circuits that operate at high speed require the multiplexer to be selected automatically. A mechanical switch cannot perform this task satisfactorily. Therefore, multiplexer used to perform high speed switching are constructed of electronic components.

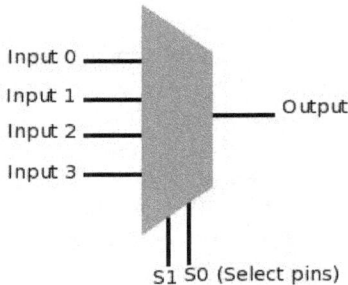

Multiplexer handle two type of data that is analog and digital. For analog application, multiplexer is built of relays and transistor switches. For digital application, they are built from standard logic gates.

The multiplexer used for digital applications, are also called **digital multiplexer**, is a circuit with many input but only one output. By applying control signals, we can steer any input to the output. Few types of multiplexer are **2-to-1, 4-to-1, 8-to-1, 16-to-1** multiplexer.

Demultiplexer

Demultiplexer means one to many. A demultiplexer is a circuit with one input and many outputs. By applying control signal, we can steer any input to the output. Few types of demultiplexer are **1-to 2, 1-to-4, 1-to-8 and 1-to 16** demultiplexer.

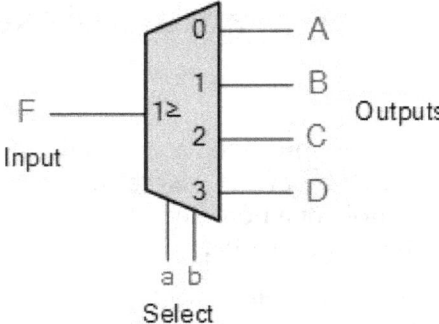

Latches and Flip flop

Latch: A latch (**bistable multivibrator**) is a device which has two stable states namely **high output** as well as **low output**. This includes a feedback lane; accordingly, data can be stored with the device. A latch is a memory device which is used to store one bit of data. These are same as flip-flops, however, they are not synchronous devices.

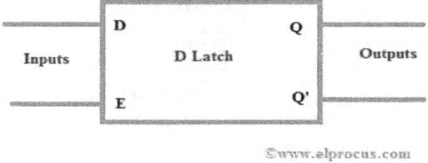

Flip flop

A **Flip-Flop** or **FF** is a couple of latches, and the designing of this can be done using a **NOR** gate or a **NAND** gate. Therefore, an FF can have **2-inputs, 2-outputs, a set as well as reset**. This type of FF is named as **SR-FF**. The main function of the flip-flop is to store the binary values. A Flip-Flop will have an extra CLK signal to make it work in a different way when contrasted with a latch.

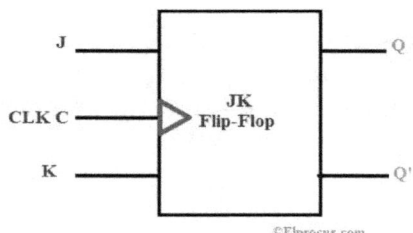

Register

A Register is a device which is used to store the information in the form of bits. It is a group of flip flops connected in series used to store multiple bits of data.

The information stored within registers can be transferred with the help of shift registers. **Shift Register** is a group of flip flops used to store multiple bits of data. The bits stored in register can move within the registers and in/out of the registers by applying clock pulses. An **n-bit** shift register can be formed by connecting n flip-flops where each flip flop stores a single bit of data.

The registers which will shift the bits to left are called *"Shift left registers"*. The registers which will shift the bits to right are called *"Shift right registers"*.

Shift registers are basically of four types:

    a. Serial in Serial Out shift register

    b. Serial in parallel Out shift register

    c. Parallel in Serial Out shift register

    d. Parallel in parallel Out shift register

CHAPTER 3
Microprocessor 8085

Basic terms

- **Hardware:** The interconnection of physical equipment to build or design a computer, like Input/Output devices, motherboard, memory, and accessories.

- **Software:** A set of instructions, data, or programs used to operate computers and execute specific tasks.

- **Firmware:** A software program or set of instructions programmed on a hardware device. It provides the necessary instructions for how the device communicates with the other computer hardware.

- **Middleware:** A software that acts as a bridge between an operating system or database and applications, especially on a network.

- **Microcomputer:** The term microcomputer is generally a synonymous with personal computer, or a computer that depends on a microprocessor.

- **Microprocessor:** A silicon chip that contains a CPU. In the world of personal computers, the terms microprocessor and CPU are used interchangeably.

- **Central Processing Unit (CPU):** The *"brain"* of a computer.
- **Memory (RAM/ROM):** A digital circuitry used to store programs and data.
- **Bus:** A multi-bit communication channel used within a computer system.
- **Bit:** It consists of 0 and 1.
- **Nibble:** Combination of 4 bits.
- **Byte:** Combination of 2 nibbles or 8 bits.
- **Word:** Combination of 2 bytes or 16 bits.
- **Double word:** Combination of 4 bytes or 32 bits.
- **Address bus:** A group wires that carries addresses of memory or I/O devices.
- **Data bus:** A group of wires that carries data from Microprocessor to memory or I/O devices and vice versa.
- **Control bus:** A group of wires that carries various control signal from Microprocessor to memory or I/O devices and vice versa.
- **Operations (op codes):** The set of basic operations that a computer can be instructed to perform, encoded in binary.
- **Operand:** The data operated on by an operation.
- **Instruction:** Combination of an op code and its operand.
- **Program:** A group of instructions that allows a computer to perform a specific job.

Pin diagram and architecture of 8085 Microprocessor

Pin diagram of 8085

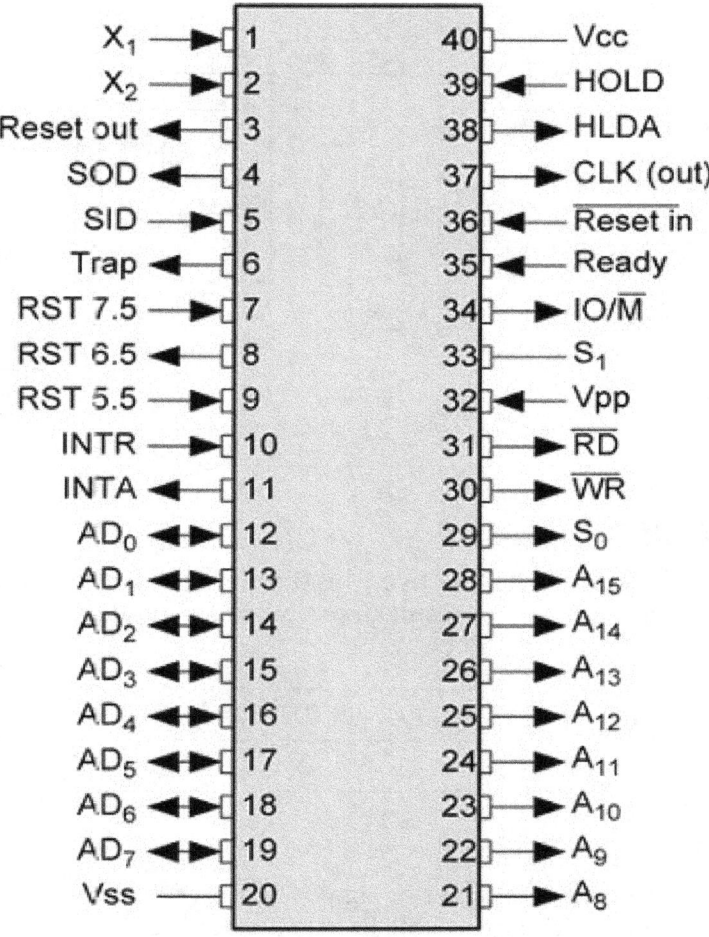

Architecture of 8085 Microprocessor

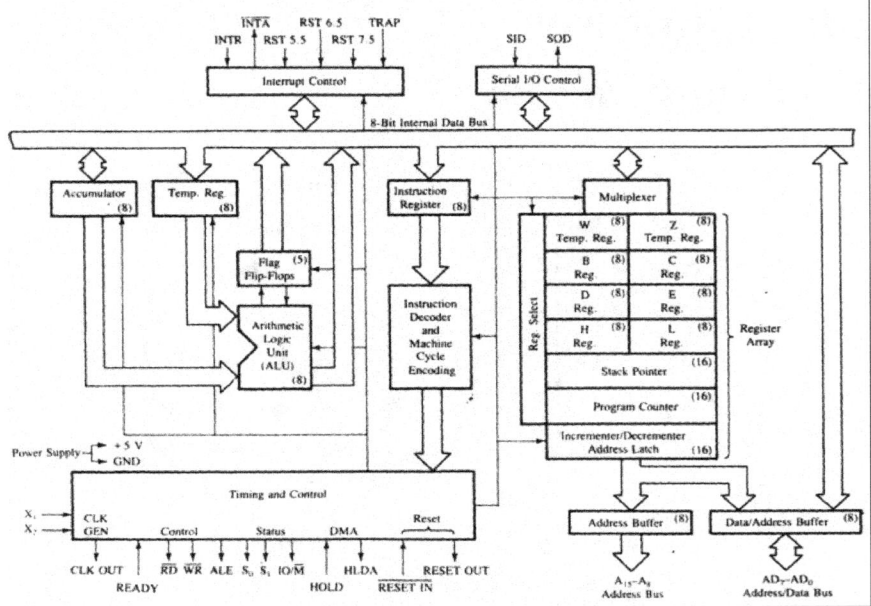

Flag register

It is an **8-bit** register having five 1-bit flip-flops, which holds either 0 or 1 depending upon the result stored in the accumulator.

Its bit position is:

D7	D6	D5	D4	D3	D2	D1	D0
S	Z	X	AC	X	P	X	CY

Sign flag (S):

It used to provide the status of sign (positive or negative) of signed number data which is available in data bus (d0 – d7).

- **If MSB bit of data is high**

    data is Negative (-ve): Sign flag will set.

- **If MSB bit of data is low**

    data is Positive (+ve): Sign flag will reset.

Zero flag (Z):

After arithmetic and logical operation

- **If answer is equal to zero:** Zero flag will set.
- **If answer is not equal to zero:** Zero flag will reset.

Auxiliary carry flag (AC):

After arithmetic and logical operation

- **If carry is generated from d3 to d4:** Auxiliary carry flag will set.
- **If carry is not generated from d3 to d4:** Auxiliary carry flag will reset.

Carry flag (CY):

After arithmetic and logical operation

- **If carry is generated from d7 to d8:** Carry flag will set.
- **If carry is not generated from d7 to d8:** Carry flag will reset.

Parity flag (P):

- **If number of one's in an accumulator is even count:** Parity flag will set.
- **If number of one's in an accumulator is odd count:** Parity flag will reset.

Interrupt

Interrupt is a signal send by an external device to the processor, to perform a task or work. In the microprocessor-based system the interrupts are used for data transfer between the peripheral and the microprocessor.

Types of Interrupt:

a. **Software Interrupt**

The software interrupts are program instructions, which are inserted at desired locations in a program.

The 8085 has eight software interrupts from RST 0 to RST 7. The vector address for these interrupts can be calculated as follows.

*Vector location = Interrupt number * 8*

Interrupt no	RST0	RST1	RST2	RST3	RST4	RST5	RST6	RST7
Vector Address/Location	0000H	0008H	0010H	0018H	0020H	0028H	0030H	0038H

b. Hardware Interrupt

An external device initiates the hardware interrupts and place an appropriate signal at the interrupt pin of the processor. If the interrupt is accepted, then the processor executes an interrupt service routine.

S. No	Name of Interrupt	Maskable	Priority	Trigger	Vectored	Vector location/ address
1.	TRAP	Non-Maskable	1st (Highest)	Edge and Level	Vectored	0024H
2.	RST 7.5	Maskable	2nd	Edge	Vectored	003CH
3.	RST 5.5	Maskable	3rd	Level	Vectored	0034H
4.	RST 6.5	Maskable	4th	Level	Vectored	002CH
5.	INTR	Maskable	5th (Lowest)	Level	Non Vectored	--

Instruction

Each instruction has two parts:

    a. First part explains the task to be performed, called the operation code (opcode)

    b. Second part tells the data to be operated on, called the operand. The operand (or data) can be specified in various ways.

The 8085-instruction set is classified into the following three groups according to word size:

    a. One-word or 1-byte instructions

    b. Two-word or 2-byte instructions

c. Three-word or 3-byte instructions

Size of instruction	Feild1	Feild2	Feild3	Length	Example
One-byte instruction	Op-code	-----	------	1 byte (8 bit)	MOV A, B
Two-byte instruction	Op-code	Operand	------	2 byte (16 bit)	MVI B,55H
Three-byte instruction	Op-code	Operand	Operand	3 byte (24 bit)	JMP 2500H

Op-code

Identification of internal register

8 Bit register	
Register	Binary Code
B	000
C	001
D	010
E	011
H	100
L	101
M	110
A	111

16 Bit register	
Register	Binary Code
BC	00
DE	01
HL	10
AF or SP	11

Examples of instruction format

S. No.	Function	Operation code							
		B7	B6	B5	B4	B3	B2	B1	B0
1	MVI R_d, data	0	0	D	D	D	1	1	0
2	LXI R_p, data	0	0	D	D	0	0	0	1
3	MOV R_d, R_s	0	1	D	D	D	S	S	S

Note:
- R_d: Destination register
- R_s: Source register
- Rp: Register pair
- DDD: Binary value of destination register
- SSS: Binary value of source register
- DD: Binary value of destination registers pair.

Addressing modes

There are different techniques called as called **addressing modes** are used to specify data for instructions.

Intel 8085 has the following addressing modes:

- **Direct Addressing mode:** In this addressing mode, the address of the operand (data) is given in the instruction itself.

  Example:

  STA 2400H: It stores the content of the accumulator in the memory location 2400H.

- **Register Addressing mode:** In register addressing mode, the operand is in one of the general-purpose registers. The opcode specifies the address of the register(s) in addition to the operation to be performed.

  Example:

  MOV A, B: Move the content of B register to register A.

- **Register Indirect Addressing mode:** In Register Indirect mode of addressing, the address of the operand is specified by a register pair.

Example:

LXI H, 2500 H: Load H-L pair with 2500H.

MOV A, M: Move the content of the memory location, whose address is in H-L pair (i.e. 2500 H) to the accumulator.

HLT: Halt.

- **Immediate Addressing mode:** In this addressing mode, the operand is specified within the instruction itself.

Example:

LXI H, 2500 is an example of immediate addressing. 2500 is 16-bit data which is given in the instruction itself. It is to be loaded into H-L pair.

- **Implicit Addressing:** There are certain instructions which operate on the content of the accumulator. Such instructions do not require the address of the operand.

Example:

CMA, RAL, RAR, etc.

Instruction set

Arithmetic Instruction

OPCODE	OPERAND	EXPLANATION	EXAMPLE
ADD	R	A = A + R	ADD B
ADD	M	A = A + Mc	ADD 2050
ADI	8-bit data	A = A + 8-bit data	ADD 50
ADC	R	A = A + R + prev. carry	ADC B
ADC	M	A = A + Mc + prev. carry	ADC 2050
ACI	8-bit data	A = A + 8-bit data + prev. carry	ACI 50
SUB	R	A = A – R	SUB B
SUB	M	A = A – Mc	SUB 2050
SUI	8-bit data	A = A – 8-bit data	SUI 50
SBB	R	A = A – R – prev. carry	SBB B
SBB	M	A = A – Mc –prev. carry	SBB 2050

OPCODE	OPERAND	EXPLANATION	EXAMPLE
SBI	8-bit data	A = A − 8-bit data − prev. carry	SBI 50
INR	R	R = R + 1	INR B
INR	M	M = Mc + 1	INR 2050
INX	r.p.	r.p. = r.p. + 1	INX H
DCR	R	R = R − 1	DCR B
DCR	M	M = Mc − 1	DCR 2050
DCX	r.p.	r.p. = r.p. − 1	DCX H
DAD	r.p.	HL = HL + r.p.	DAD H

Logical instructions

OPCODE	OPERAND	DESTINATION	EXAMPLE
ANA	R	A = A AND R	ANA B
ANA	M	A = A AND Mc	ANA 2050
ANI	8-bit data	A = A AND 8-bit data	ANI 50
ORA	R	A = A OR R	ORA B
ORA	M	A = A OR Mc	ORA 2050
ORI	8-bit data	A = A OR 8-bit data	ORI 50
XRA	R	A = A XOR R	XRA B
XRA	M	A = A XOR Mc	XRA 2050
XRI	8-bit data	A = A XOR 8-bit data	XRI 50
CMA	None	A = 1's compliment of A	CMA
CMP	R	Compares R with A and triggers the flag register	CMP B
CMP	M	Compares Mc with A and triggers the flag register	CMP 2050
CPI	8-bit data	Compares 8-bit data with A and triggers the flag register	CPI 50
RRC	None	Rotate accumulator right without carry	RRC

OPCODE	OPERAND	DESTINATION	EXAMPLE
RLC	None	Rotate accumulator left without carry	RLC
RAR	None	Rotate accumulator right with carry	RAR
RAL	None	Rotate accumulator left with carry	RAR
CMC	None	Compliments the carry flag	CMC
STC	None	Sets the carry flag	STC

Interfacing

Schematic Block Diagram for Memory and I/O interfacing with Microprocessor:

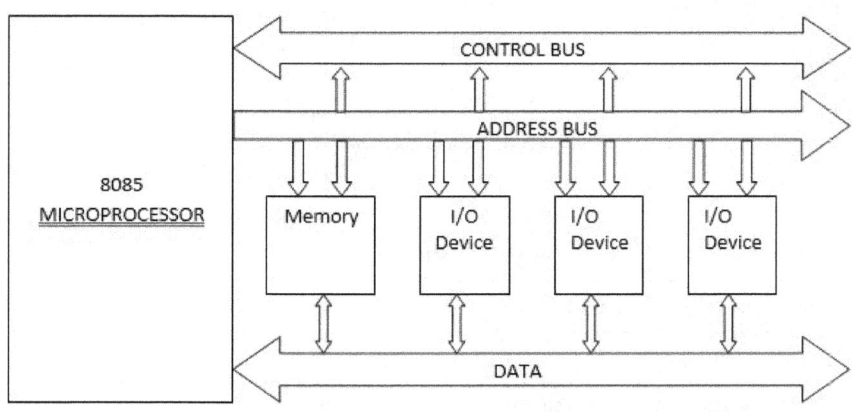

Schematic Block Diagram of I/O Mapped I/O Devices:

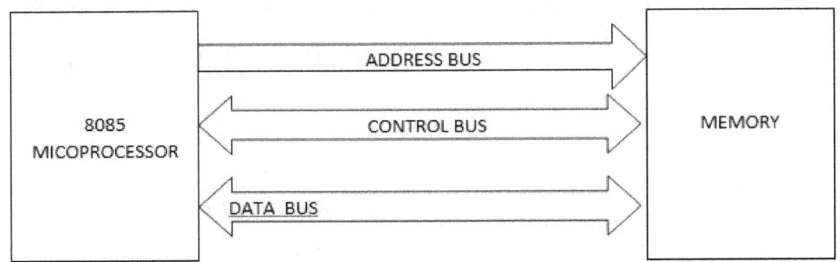

Schematic Block Diagram of Memory Mapped I/O scheme:

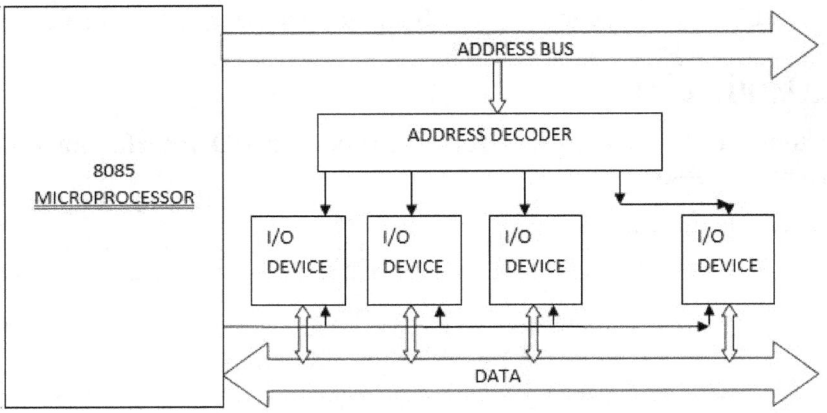

Memory Mapped I/O vs. I/O Mapped I/O

Memory-Mapped I/O	I/O Mapped I/O
16-bit Address	8-bit Address
Memory read and memory write signals	I/O read and I/O write signals
Memory related instructions MOV M, R, MOV R, M, ADD M ANA M, STA, LDA, STAX, LDAX	I/O related instructions IN and OUT
Data transfer is between any register and I/O	Data transfer is between accumulator and I/O
More hardware needed to decode 16-bit address	Less hardware is used to decode 8-bit address

The memory map 64K is shared between I/Os and system memory	The I/O mapped is independent of the memory map; 256 I/O devices
Arithmetic and logical operations can be directly performed with I/O data.	Not available

Schematic Block Diagram of Input Data:

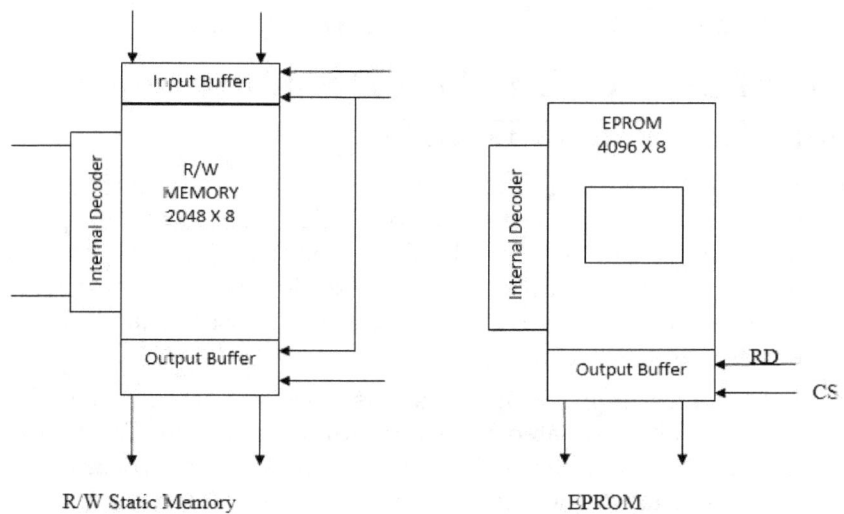

R/W Static Memory　　　　　　　EPROM

Example 3.1: Interfacing of 4K EPROM with 8085 with starting address A000H

Step1: Identify the number of address lines and Chip select from the given memory size.

> Here, memory size is 4K or 4096 bytes. Therefore, memory chip requires 12 address lines from A0-A11 to decode 4096-byte locations. The remaining address lines A12-A15 are connected to decoder to generate Chip Select signal.

Step 2: Draw the memory address allocation table

The logic levels on the address lines A0-A11 can assume any combination from all 0s to all 1s

A15	A14	A13	A12	A11	A10	A9	A8	A7	A6	A5	A4	A3	A2	A1	A0	
1	0	1	0	0	0	0	0	0	0	0	0	0	0	0	0	Start Address
1	0	1	0	1	1	1	1	1	1	1	1	1	1	1	1	End Address

So, the end address is AFFFH. The values for A15 to A12 are according to the decoding circuit that remains same for both the start address and end address.

Step 3: Draw the interfacing circuit of 8085 with memory.

Address Decoding Techniques

Absolute decoding/Full Decoding Linear decoding/Partial Decoding

- **Absolute decoding**: In absolute decoding technique, all the higher address lines are decoded to select the memory chip, and the memory chip is selected only for the specified logic levels on these high-order address lines; no other logic levels can select the chip. This addressing technique is normally used in large memory systems.

- **Linear decoding:** In small systems, hardware for the decoding logic can be eliminated by using individual high-order address lines to select memory chips. This is referred to as **linear decoding** or **partial decoding**. It reduces the cost of decoding circuit, but it has a drawback of multiple addresses (shadow addresses).

  A15 address line is directly connected to the chip select signal of **EPROM** and after inversion it is connected to the chip select signal of the RAM. Therefore, when the status of A15 line is *'zero'*, EPROM gets selected and when the status of A15 line is *'one'* RAM gets selected. The status of the other address lines is not considered, since those address lines are not used for generation of chip select signals.

  A0-A11 address lines are directly connected to address bus of memory chip. A12-A15 are used for generating chip select signal for memory chip.

  Address decoding circuit using 3X8 decoder:

  A15 line is use for enabling 74LS138 decoder chip. A12, A13, A14 lines are connected to 74LS138 chip as inputs. When these lines are 010 output should be '0'. This is provided at O2 pin of 74LS138 chip.

Assembly language program

- **Compiler**

  Compiler is the language processor that reads the complete source program written in high level language as a whole in one go and translates it into an equivalent program in machine language.

  Example: C, C++, C#, Java

- **Assembler**

  Assembler is used to translate the program written in Assembly language into machine code. The source program is an input of assembler that contains assembly language instructions. The output generated by assembler is the object code or machine code understandable by the computer.

- **Interpreter**

  The translation of single statement of source program into machine code is done by language processor and executes it immediately before moving on to the next line is called an interpreter. If there is an error in the statement, the interpreter terminates its translating process at that statement and displays an error message. The interpreter moves on to the next line for execution only after removal of the error. An Interpreter directly executes instructions written in a programming or scripting language without previously converting them to an object code or machine code.

  Example: Perl, Python, and Matlab.

Assembler directives of 8085

An assembler directive is a message or instruction to the assembler to know the assembly process. For example, an assemble directive tests the assembler where a program is to be in memory.

- **EQU Equate:** Assigns a value to a symbol (same as =)
- **ORG Origin:** It sets the current origin to a new value. This is used to set the program or register address during assembly. For example, ORG 0100h tells the assembler to assemble all subsequent code starting at address 0100h.

- **DC Define constant:** Define constant assembler directive allows you to put a data value in the memory at the time when the program is first loaded.
- **DS Define storage:** It defines an amount of free space. No code is generated and is sometimes used for allocating variable space.
- **END:** End of assembly language program and *"starting address"* for execution.

Delay routine

In delay routine a count (number) is loaded in a register of microprocessor. Then it is decremented by one and the zero flag is checked to verify whether the content of the register is zero or not. This process is continued until the content of register is zero. When it is zero, the time delay is over, and the control is transferred to main program to carry out the desired operation.

Consider an example, if the 8085 microprocessor has 5 MHz quartz crystal then, the internal clock frequency = 5 / 2 = 2.5 MHz Time for one T-state= 1 / 2.5 x 10⁶ = 0.4μsec

For small time delays (< 0.5 msec) an 8- bit register can be used.

For large time delays (< 0.5 Sec) 16-bit register should be used.

For very large time delays (> 0.5 sec), a delay routine can be repeatedly called in the **main program**.

CHAPTER 4
Peripheral Devices and Interfacing

Programmable peripheral Interface (8255)

Programmable peripheral Interface (PPI) 8255 is a general-purpose programmable device designed to interface the CPU with its I/O like ADC, DAC, keyboard, and so on. It can be used with microprocessor.

It consists of three 8-bit bidirectional I/O ports i.e. PORT A, PORT B and PORT C. Ports can be assigned as input or output functions.

Block diagram of 8255

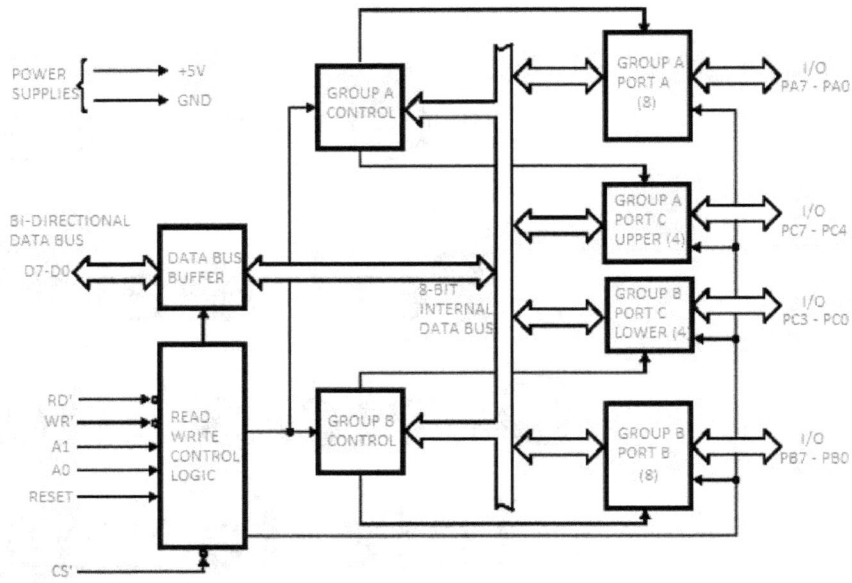

Figure 4.1: Block diagram of 8255

8285 consists of 40 pins and operates in +5V regulated power supply. Port C is further divided into two 4-bit ports i.e. port C lower and port C upper and port C can work in either **bit set rest (BSR)** mode or in mode 0 of input-output mode of 8255. Port B can work in either mode 0 or in mode 1 of input-output mode. Port A can work either in mode 0, mode 1, or mode 2 of input-output mode. It has two control groups; control group A and control group B. Control group A consist of port A and port C upper. Control group B consists of port C lower and port B. Depending upon the value of CS', A1 and A0 we can select different ports in different modes as input-output function or BSR. This is done by writing a suitable word in control register (control word D0-D7).

Programmable interval timer (8253/8254)

8254 is a device designed to solve the timing control problems in a microprocessor. It has 3 independent counters, each capable of handling clock inputs up to 10 MHz and size of each counter is 16-bit. It operates in +5V regulated power supply and has 24 pin signals. All modes are software programmable. The 8254 is an advanced version of 8253 which did not offer the feature of read back command.

Block diagram of 8253/54

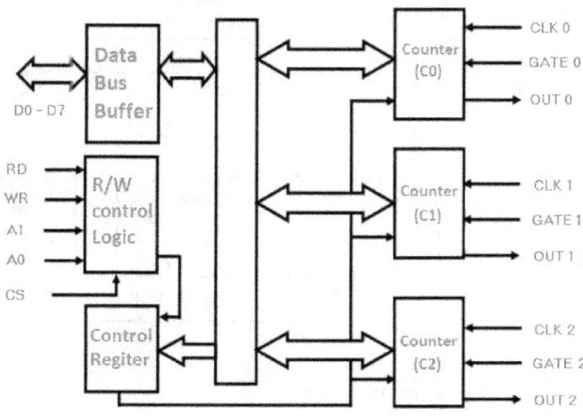

Figure 4.2: Block diagram of 8253/54

It has 3 counters each with two inputs (Clock and Gate) and one output. Gate is used to enable or disable the counting. When any value of count is loaded and value of gate is set (1), after every step value of count is decremented by 1 until it becomes zero. Depending upon the value of CS, A1, and A0 we can determine the addresses of selected counter.

Programmable interrupt controller (8259)

The 8259A is a *"programmable interrupt controller"* designed to work with microprocessor like 8080, 8085A, 8086, 8088.

PIC can handle 08 interrupt inputs. 8259 can vector an interrupt request anywhere in the memory in 8085A microprocessor. However, all the 08 interrupts are spaced at a regular interval of either 04 or 08 locations.

Block diagram of 8259

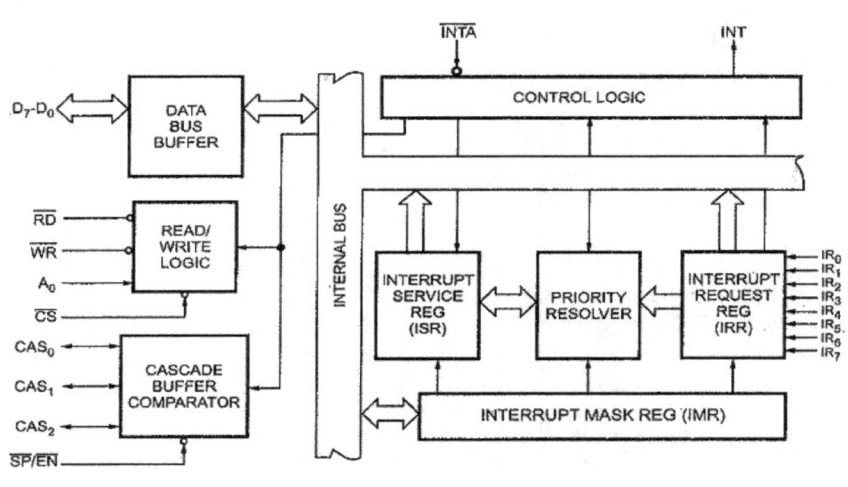

Fig. 14.71 Block diagram of 8259A

Figure 4.3: Block diagram of 8259

8259 can be programmed to an accept interrupt requests either as edge triggered interrupt or level triggered or request unlike your RST interrupts where some are edge triggered and some are level triggered. It can resolve 08 levels of interrupt priorities in variety of modes. Lower priority interrupts can be allowed to be acknowledging during the service of higher priority interrupts.

Direct Memory Access controller (8257)

DMA stands for *"Direct Memory Access"*. It is designed to transfer data at fastest rate. DMA allows transfer of data directly from/to memory without any CPU interference.

The device requests the CPU using a DMA controller to hold its address, data, and control bus, so the device becomes free to transfer the data directly. Data transfer can be initiated only after receiving HLDA signal from the CPU.

Block diagram of 8257

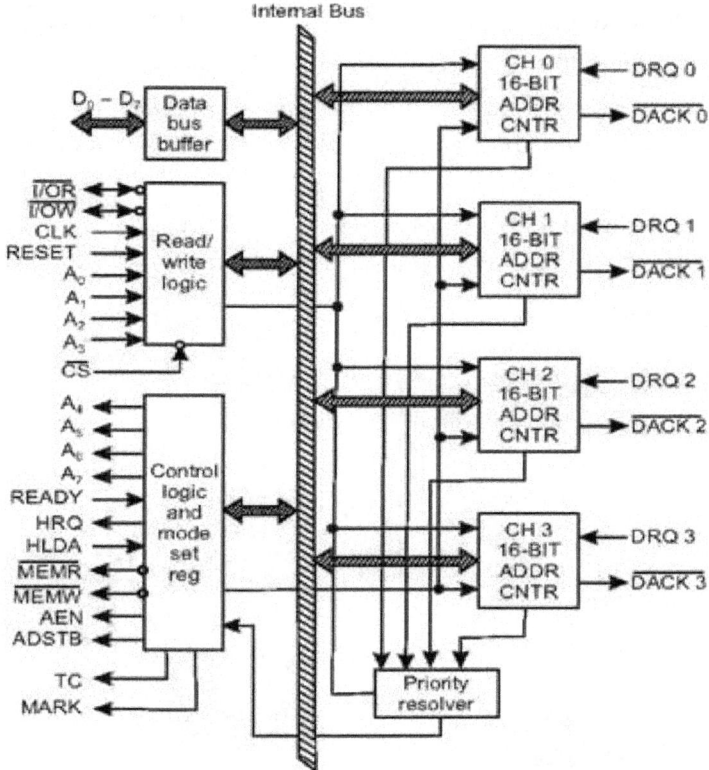

Figure 4.4: Block diagram of 8257

There are four channels which can be used for four I/O devices. Each channel has 14-bit counter and 16-bit address. Each channel can be programmed independently.

Each channel can perform write transfer, read transfer, and verify transfer operations. It requires a single-phase clock. It operates in 2 modes, i.e., Master mode and Slave mode. Its frequency ranges from 250Hz to 3MHz.

Universal Synchronous Asynchronous Receiver Transmitter (8251)

Universal synchronous asynchronous receiver transmitter (USART) acts as a mediator between microprocessor and peripheral to transmit serial data into parallel form and vice versa. It takes data serially from peripheral (outside devices) and converts into parallel data. After converting the data into parallel form, it transmits it to the CPU.

Similarly, it receives parallel data from microprocessor and converts it into serial form. After converting data into serial form, it transmits it to outside device (peripheral).

Block diagram of 8251

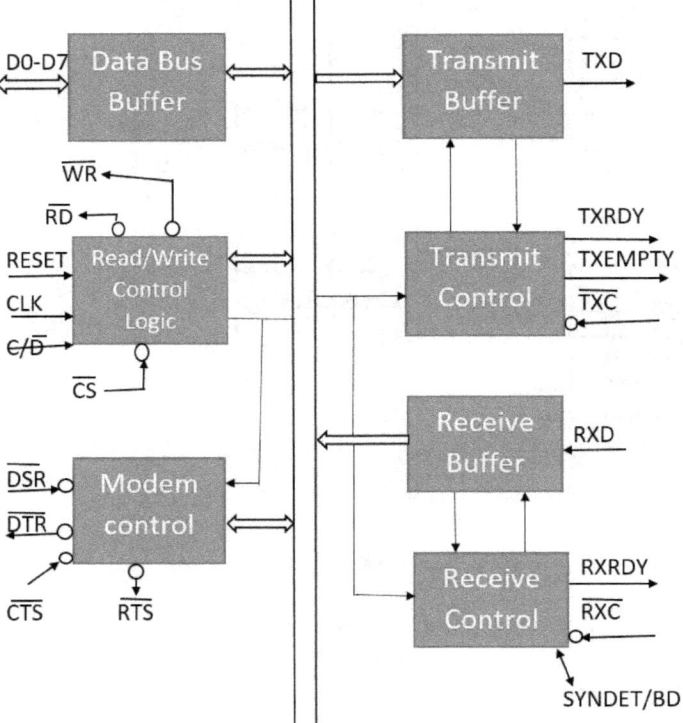

Figure 4.5: Block diagram of 8251

CHAPTER 5
AVR ATmega32

Introduction

AVR microcontroller is the latest microcontroller developed by **Atmel** in **1986**. AVR is just a product line from Atmel but many times AVR might stands for **advanced virtual RISC** or **Alf** and **Vegard RISC**. AVR is RISC based architecture machine with various types of microcontrollers like 32-bit microcontroller, 8-bit microcontroller. AVR supports various inbuilt features like ADC, PWM, USART, SPI, I2C, CAN, and so on. 8-bit AVR microcontroller contains minimum of 32 general purpose resistors which all are capable to operate as Accumulator. Thus, AVR microcontroller does not have any dedicated accumulator resistor, all its general-purpose resistors are capable to act as accumulator.

Features of ATmega32 is a 40-pin IC:

- 32 general-purpose resistors
- 2144 bytes of data memory with 64 bytes of I/O resistor, 2048 bytes of ASRAM and 32 general-purpose resistors
- 32 Kbyte of code memory
- 1 Kbyte of EEPROM
- 32 I/O pins
- 8 channels 10-bit ADC
- 3 timers

- SPI bus
- I2C bus

The pin diagram of ATmega32

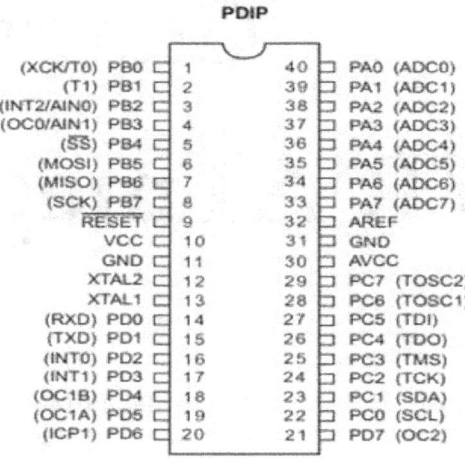

Figure 5.1: Pin diagram of ATmega32

ATmega32 flash is of 32 Kbytes, which is arranged as 16kX16 and its program counter is 14-bit wide.

It has 5 ports named as PORT A, PORT B, PORT C, PORT D, PORT E.

- PORTA is 8-bit I/O PORT and supports 8 channel 10-bit ADC. On RESET it act as a I/O PORT but through programming it can be made to act as analog input for inbuilt ADC.
- PORTB, PORTC, and PORTD are 8-bit I/O PORT with other programmable function supported like SPI, I2C, timer input, interrupt input, USART.

For programming and reading/writing the data to I/O PORTs 3 resistors involved are:

- DDRx
- PORTx
- PINx

Where 'x; indicates the PORT

DDR is a data direction resistor and PIN is PORT input pins. DDR resistor is used to define a PORT as input or output.

ATmega32 has 3 timers:
- Timer0
- Timer1
- Timer2

Timer0 and Timer2 are 8-bit Timers, while Timer1 is 16-bit Timer. Timer can be used for generation of delay by using the internal clock pulse and if Timer is fed with external clock pulse it is called **counter**. The resistors/flags which are involved in the programming of Timers are:

- **TCNTn-Timer/counter resistor:** This resistor is loaded with the value and counts up for each clock pulse till it overflows.
- **TOVn-Timer overflow flag:** When the Timer gets the overflows, this flag is set.
- **TCCRn-Timer/counter control resistor:** This resistor is used for setting up modes of Timer resistor.
- **OCRn-output compare resistor:** The content of the resistor is compared with TCNTn resistor and when both are equal the flag OCFn-output compare flag is set.

AVR supports communication full duplex USART communication, the major resistor which are involved in serial communication are:

- UDR-USART data resistor
- UCSRA, UCSRB, UCSRC-USART control resistor
- UBRR-USART baud rate resistor

UBBR resistor sets the baud rate for serial communication as per following formula:

Desired baud rate = $F_{osc}/(16(X+1))$

Where 'X' is value to be loaded UBBR resistor

ATmega32 has 10-bit 8 channel ADC which is available along with PORTA programmable pin. The main characteristics of ADCs are:

- **Resolution:** This is also called as **step size** which is a smallest change that can be detected by ADC in the input analog voltage. The step size or resolution depends on the bit size of the ADC. 16-bit ADC will be having higher resolution than 10-bit ADC
- **Conversion time**: Conversion time is the time which the ADC takes to convert the analog signal into digital. It is dictated by the clock source connected to ADC
- **Vref:** It is the input reference voltage for ADC, the voltage connected to this pin along with the resolution of the ADC defines the step size.

Characteristics of ATmega32 ADC:

- The ADC is of 10-bit, ADC in ATmega32 is with 8 channel analog input with 7 differential input channels and 2 channels with optional gain of 10x and 200x.
- The converted digital data is stored in ADCL (A/D result low) ADCH(A/D result high). ADCH: ADCL is of 16 bit and digital data is of 10-bit. Thus, it gives the option of making either upper 6-bits or lower 6-bits as unused.
- ATmega32 gives 3 options for Vref, it can be connected to AVCC, internal 2.56 volt or external AREF pin.
- The conversion time for ATmega32 ADC is calculated by the crystal oscillator connected to XTAL pins and ADPS0:2 bits of ADCSRA (A/D control and status resistor A)
- The ADC channel for ATmega32 is selected with MUX0:4 bits of ADMUX resistor
- Vref source is selected through REFS0:1 bit of ADMUX resistor.

Chapter 6

Interfacing of Input/Output Device

Light Emitting Diode (LED)

LED is semiconductor device and stands for Light emitting diode. LED is interface with micro controller with for displaying the binary information. LED has two terminals as anode and cathode. When LED anode is connected to positive supply and LED cathode is connected to ground LED gets ON and on reverse connection LED gets OFF

Seven Segment

Seven segments are the type of display used for displaying the numeric data and to some extent alphabets also. Seven segment display is made of LEDs which are arranged in the shape of EIGHT. There are two types of seven segments

a. Common cathode (CC)
b. Common anode (CA)

The difference between two types of segment displays is depicted in the *figure 6.1* along with the pin diagram of seven segment device.

Each LED in seven segment display is name from **a** to **g** and **dp** sometime **h**. In order to display the numerical character on seven segments, programmer is required to create 8-bit data for the level i.e. HIGH or LOW of each segment in seven segment display from **a** to **h (dp)**. The table for common anode and common cathode to display from 0 to nine is:

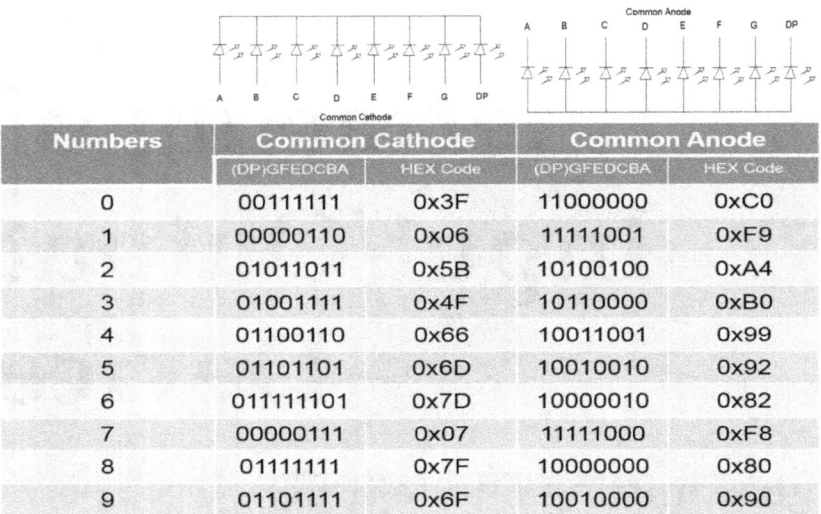

Numbers	Common Cathode		Common Anode	
	(DP)GFEDCBA	HEX Code	(DP)GFEDCBA	HEX Code
0	00111111	0x3F	11000000	0xC0
1	00000110	0x06	11111001	0xF9
2	01011011	0x5B	10100100	0xA4
3	01001111	0x4F	10110000	0xB0
4	01100110	0x66	10011001	0x99
5	01101101	0x6D	10010010	0x92
6	011111101	0x7D	10000010	0x82
7	00000111	0x07	11111000	0xF8
8	01111111	0x7F	10000000	0x80
9	01101111	0x6F	10010000	0x90

Table 6.1: Common anode and Common cathode

Liquid Crystal Display

Liquid Crystal Display is used for alphanumeric display. LCD is also capable of displaying some special characters with programming. 16x2 is a basic character LCD used to display alphanumeric information.
Feature of 16x2 LCD:
- 16x2 has 32-character space
- Each character space is of 5x8 pixel size
- 5x7 pixel area is used for character display
- 8th row is reserved for cursor

Pin diagram of 16x2 LCD

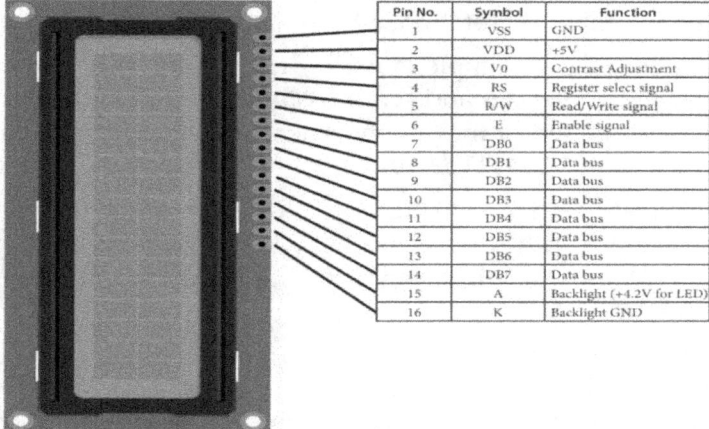

Pin No.	Symbol	Function
1	VSS	GND
2	VDD	+5V
3	V0	Contrast Adjustment
4	RS	Register select signal
5	R/W	Read/Write signal
6	E	Enable signal
7	DB0	Data bus
8	DB1	Data bus
9	DB2	Data bus
10	DB3	Data bus
11	DB4	Data bus
12	DB5	Data bus
13	DB6	Data bus
14	DB7	Data bus
15	A	Backlight (+4.2V for LED)
16	K	Backlight GND

Figure 6.1: Pin diagram of 16x2 LCD

Commands used for LCD initialization are:

LCD Command Codes

Code (Hex)	Command to LCD Instruction Register
1	Clear display screen
2	Return home
4	Decrement cursor (shift cursor to left)
6	Increment cursor (shift cursor to right)
5	Shift display right
7	Shift display left
8	Display off, cursor off
A	Display off, cursor on
C	Display on, cursor off
E	Display on, cursor blinking
F	Display on, cursor blinking
10	Shift cursor position to left
14	Shift cursor position to right
18	Shift the entire display to the left
1C	Shift the entire display to the right
80	Force cursor to beginning to 1st line
C0	Force cursor to beginning to 2nd line
38	2 lines and 5x7 matrix

Motors

Motors are the electromechanical devices which converts the electrical energy into mechanical rotation. **Different types of motors are**:

a. **DC motors** are the simple motors which rotated when excitation is given to it. Stepper and servo motors are used in applications where precise motor movement is needed to be controlled like in robotic or in CD/DVD drives.

   DC and stepper motor required motor drive device to interface these motors with microcontroller. While servo motor requires **the PWM – Pulse width modulation signal** for controlling the motor.

   DC motor has only two terminals. In order to start the motor power supply as per the rating of the motor is provided to the terminals of motor with one terminal connected to positive supply and second terminal to ground. By changing the power supply connections to the motor terminals direction of rotation of DC motor can be change. DC motor speed is controlled by varying the applied power supply to the motor within allowed range. L293d is the generic DC motor driver IC used to interface the DC motor with microcontroller.

b. **Stepper motor** has 4 terminals generally named as A, B, C, D. By providing the predefined sequence to the motor terminals motor is rotated with precise stepping. By sending the sequence of signal the stepper motor rotates in one direction and in order to revert the rotation direction the sequence of signal need to be reversed. **ULN2003A** is the generic stepper motor drive used to interface stepper motor with microcontroller.

c. **Servo motor** is generally used in robotics and has three terminal two for power supply and the third for PWM signal. By providing the pulse of required width the angle of rotation in controlled for servo motor.

Exercise

Descriptive Type Questions

1. What is a Logic gate?

Answer: Logic gates are the basic elements to make up a digital system. The electronic gate is a circuit that can operate on a number of binary inputs in order to perform a particular logicaa function.

2. Name the basic logic operators.

Answer:

The basic logic operators:

- NOT / INVERT
- AND
- OR

3. Give the classification of logic families

Answer:

Classification of logic families are:

- Bipolar Unipolar
- Saturated, Non-Saturated PMOS NMOS
- CMOS
- RTL Schottky TTL ECL DTL
- I I C TTL

4. Which gates are called as the universal gates? What are its advantages?

Answer: The NAND and NOR gates are called as the universal gates. These gates can be used to perform any type of logic application.

5. Explain why a two-input NAND gate called universal gate?

Answer: NAND gate is called universal gate because any digital system can be implemented with the NAND gate. Sequential and combinational circuits can be designed with these gates because element circuits like flip-flop can be constructed from two NAND gates connected back-to-back. NAND gates are common in hardware because they are easily available in the ICs form.

6. What are the types of computers?

Answer:

Types of computer are:

a. Supercomputer

b. Main frame computer

c. Minicomputer

d. Microcomputer

7. Draw the basic block diagram of Microcomputer system.

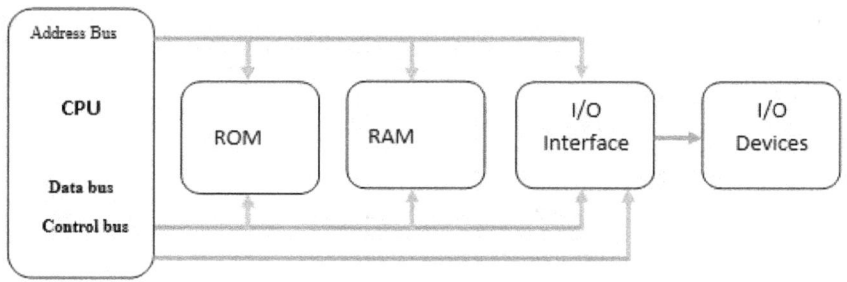

8. Discuss the types of computer architecture.

Answer:

The types of computer architecture are:

- **Princeton or von-Neumann architecture**
o   It has a single memory which is used for both code and data.

o It requires two clock cycles, first clock cycle for collecting the code and second clock cycle for collecting the data.
- **Harvard Architecture**
o It has separate memories for code and data.
o It requires only one clock cycle for collecting code and data.

9. What are the major components of microcomputer systems?

Answer:

The major components of microcomputer systems are:
- CPU
- Memory
- I/O devices
- Clock generator

10. What are the three main functions of CPU?

Answer:

The major functions of CPU are:
- Fetch
- Decode
- Execute

11. What is a Microprocessor?

Answer: Microprocessor is a program-controlled device, which fetches the instructions from memory; then decodes it and executes the instructions. Most of the Micro Processors are single- chip devices.

12. Give the example of Hardware Interrupts.

Answer: TRAP, RST7.5, RST6.5, RST5.5 and INTR.

13. Name the level-triggering Interrupt?

Answer: RST 6.5 and RST 5.5 are the level-triggering interrupts.

14. Classify the signals for 8085 microprocessor?

Answer:

The signals for 8085 microprocessor are:
- Frequency and power signals
- Address and data buses
- The control bus
- Interrupt Signals
- Serial Input / Output signals
- DMA signals
- Reset Signals

15. Can RC circuit be used as clock source for 8085?

Answer: Yes, RC circuit can be used as clock source for 8085, if an accurate clock frequency is not required. Also, the component cost is low as compared to LC or Crystal.

16. Which interrupt has the highest priority?

Answer: TRAP interrupt has the highest priority

17. What is the RST for the TRAP?

Answer: RST 4.5 is called as TRAP.

18. Why crystal is preferred for clock source?

Answer: The main reason to use crystal is high stability, large Q factor and accurate frequency which remain constant all the time.

19. What is the significance of the HOLD and HLDA pins?

Answer:

- **HOLD:** It indicates that another device is requesting the use of the address and data bus. Receiving HOLD request the microprocessor relinquishes the use of the buses as soon as the current machine cycle is completed. Internal processing may continue. After the removal of the HOLD signal the processor regains the bus.

- **HLDA:** It is a signal which indicates that the HOLD request has been received and after the removal of a HOLD request, the HLDA goes low.

20. The Input/Output signals are related to which pins?

Answer: IO/M: It is a status signal which determines whether the address is for input-output or memory. When it is high (1), the address on the address bus is for input-output devices. When it is low(0) the address on the address bus is for the memory.

21. In 8085, power and frequency can be checked by connecting the wire with which pins?

Answer: Power and frequency can be checked by connecting wire with VCC, GND, X1, and X2 pins of 8085 Microprocessor.

22. How many bits address bus does the 8085 has?

Answer: 8085 has 16-bit address, by using 16-bit address 8085 Microprocessor can address 65,536 different memory locations.

23. What is the maximum clock frequency used by the 8085 microprocessor?

The maximum clock frequency of 8085 microprocessor is 3 MHz and it is called **processor clock**. To achieve this clock frequency a 6 MHz needs to be connected on X1 and X2 pins. External clock is divided by 2 in internal architecture.

24. What is the purpose of the ALE pin?

Answer: Address latch enable (ALE) is used to separate the address and data bus. The address bus is enabled during the 1st clock cycle as the ALE pin goes high. During 2nd and 3rd clock cycles it goes low, indicating the address & data bus (AD0-AD7) is for data.

25. Mention the purpose of SID and SOD lines.

Answer:

- **Serial input data line (SID):** It is an input line through which the microprocessor accepts serial data.
- **Serial output data line (SOD):** It is an output line through which the microprocessor sends output serial data.

26. Give examples for 8 / 16 / 32-bit Microprocessor?

Answer:

Examples of 8/16/32-bit microprocessor are:

- 8-bit Processor - 8085 / Z80 / 6800;
- 16-bit Processor - 8086 / 68000 / Z8000;
- 32-bit Processor - 80386 / 80486

27. What is Tri-state logic?

Answer: Three Logic Levels are used, and they are 'High', 'Low', 'High impedance' state. The high and low are normal logic levels & high impedance state is electrical open circuit conditions. Tri-state logic has a third line called enable line.

28. What is the role of s0 and s1?

Answer: s0 and s1 are the output status signals which is used to give information of operation performed by microprocessor. The S0 and S1 lines specify four different conditions of 8085 machine cycles.

S1	S0	Operation
0	0	Halt
0	1	Write
1	0	Read
1	1	Op-code fetch

29. What are the various registers in 8085?

Answer: Accumulator register, Temporary register, Instruction register, Stack Pointer, Program Counter are the various registers in 8085.

30. What is stack pointer?

Answer: Stack pointer is a special purpose 16-bit register in the Microprocessor, which holds the address of the top of the stack.

31. What is program counter?

Answer: Program counter holds the address of either the first byte of the next instruction to be fetched for execution or the address of the next byte of a multi byte instruction, which has not been completely fetched.

In both the cases it gets incremented automatically one by one as the instruction bytes get fetched. Also, Program register keeps the address of the next instruction.

32. Which stack is used in 8085?

Answer: Last In First Out (LIFO) stack is used in 8085. In this type of Stack the last stored information can be retrieved first.

33. Name five addressing modes of 8085.

Answer: Five addressing modes of 8085 are: Immediate, Direct, Register, Register indirect, Implicit addressing modes.

34. Name few Input & Output Devices?

Answer: Keyboards, Floppy disk are the examples of input devices. Printer, LED / LCD display, CRT Monitor are the examples of output devices.

35. Mention the various functional blocks of 8085 microprocessor.

Answer:

The various functional blocks of the 8085 microprocessor are:

- Registers
- Arithmetic logic unit
- Address buffer
- Increment / decrement address latch
- Interrupt control
- Serial I/O control
- Timing and control circuitry
- Instructions decoder and machine cycle encoder.

36. Mention the steps in the interrupt driven mode of data transfer.

Answer:

The steps followed in the interrupt driven mode of transfer are:

- The peripheral device would request for an interrupt.
- The request acknowledgement for the transfer is issued at the end of instruction execution.
- Now, the ISS routine is initialized, The PC has the return address which is now stored in the stack.

- Data transfer is managed and coordinates by the ISS.
- Again, the Interrupt system is enabled, and the above steps are repeated.

37. Explain briefly what happens when the INTR signal goes high in 8085.

Answer: The INTR is a maskable interrupt for the 8085. It has the lowest priority and is also non vectored. When this INTR signal goes into the high state the following actions take place:

- Every instruction executed by the 8085 checks the status of INTR interrupt.
- Till an instruction is completed the signal of INTR will remain high. Once an instruction is completed the processor sends an acknowledgement signal INTA.
- As soon as the INTA signal goes low a new opcode is placed on the data bus for transfer.
- Once the new instruction is received the processor saves the address of new instruction into the STACK and an interrupt service subroutine begins.

38. Explain the control and timing circuitry of 8085.

Answer:

The control and timing circuitry of 8085 are:

- The timing and control circuitry section of the 8085 is responsible for the generation of timing and control signals so that instructions can be executed.
- The types of signals involved are: Clock signals, Control signals, Status signals, DMA signals and the reset section.
- It is responsible for the fetching and the decoding of the various operations.
- This section also aids in the generations of control signals for the executions of instructions and for the sync between external devices.

39. What is a stack pointer register?

- The stack pointer is a sixteen-bit register used to point at the stack.
- In read write memory the locations at which temporary data and return addresses are stored is known as the **stack**.
- In simple words stack acts like an auto decrement facility in the system.
- The initialization of the stack top is done with the help of an instruction LXI SP.
- In order to avoid program crashes; a program should always be written at one end and initialized at the other.

40. Discuss the accumulator register of 8085.

Answer:

Features of accumulator registers of 8085 are:

- It is one of the most important 8-bit register of 8085.
- It is responsible for coordinating input and output to and from the microprocessor through it.
- The primary purpose of this register is to store temporary data and for the placement of final values of arithmetic and logical operations.
- This accumulator register is mainly used for arithmetic, logical, store and rotate operations.

41. In 8085 name the 16-bit registers?

Answer: BC, DE, HL, Stack pointer and Program counter.

42. What is the purpose of instruction decoder?

Answer: Instruction decoder is an 8-bit register. When an instruction is fetched from the memory it is stored in the Instruction register. Instruction decoder decodes the information present in the instruction register.

43. What are the ways to classify control and timing signals?

Answer: Control and timing signals provides timing and control signal to the microprocessor to perform operations. Following are the timing and control signals, which control external and internal circuits:

- **Control Signals:** READY, RD', WR', ALE
- **Status Signals:** S0, S1, IO/M'
- **DMA Signals:** HOLD, HLDA
- **RESET Signals:** RESET IN, RESET OUT

44. What is the need of temporary register?

Answer: Temporary register is an 8-bit nonprogrammable resister, which is used to hold data during an arithmetic and logic operation (temporary resister is used to hold intermediate result). The result is stored in the accumulator, and the flags (flip-flops) are set or reset according to the result of the operation.

45. Classify the instruction set based on their operation.

Answer:

Instruction set based on their operation are:
- Data transfer group
- Arithmetic group
- Logical group
- Branch control group
- I/O and machine control group

46. Define mnemonics.

Answer: Mnemonics are the short form of describing instructions. The mnemonics are given by the manufacturers of microprocessor and programmable devices.

47. What is a processor cycle (Machine cycle)?

Answer: The processor cycle or machine cycle is the basic external operation performed by the processor. To execute an instruction, the processor will run one or more machine cycles in an order.

48. What is instruction cycle?

Answer: The sequence of operation that a processor must carry out while executing an instruction is called instruction cycle. Each instruction cycle of a processor in turn consist of several machine cycles.

49. What is fetch and execute cycle?

Answer: The instruction cycle is consisting of fetch cycle and execution cycle. During fetch cycle processor reads the opcode from the memory and during execute cycle processor executes the instruction which is in decoded form from the instruction decoder.

50. List the various machine cycles of 8085.

Answer:

The various machine cycles of 8085 are:
- Opcode fetch cycle
- Memory read cycle
- Memory write cycle
- I/O read cycle
- I/O write cycle
- Interrupt acknowledge cycle
- Bus idle cycle

51. What is T-state?

Answer: T-state: One sub-division of an operation performed in one clock cycle is called a T-state.

52. Define opcode and operand.

Answer: An opcode is a single instruction that can be executed by the CPU. In machine language, it is a binary or hexadecimal value such as 'B6' loaded into the instruction register.

In assembly language mnemonic form an opcode is a command such as MOV or ADD or JMP.

For example

  MVI A, 34H

  The opcode is the MVI instruction. The other parts are called the *'operands'*.

Operands are manipulated by the opcode. In this example, the operands are the register named A and the value 34 hex.

53. How the 8085 processor differentiate memory access (read/write) and IO access (read/write)?

Answer: In 8085 processor memory access and I/O access is differentiate by the help of IO/$\overline{M}\overline{M}$. When IO/$\overline{M}\overline{M}$ = 0, then Processor will access or do the memory operation whereas IO/$\overline{M}\overline{M}$ = 1, then processor will access or do IO operation.

54. When does the 8085 processor checks for an interrupt?

Answer: While execution of program (set of instruction), in the second T-state of the last machine cycle of every instruction, the processor checks whether an interrupt request is made or not.

55. What will be the status of the processor during bus idle cycle?

Answer:

The status of the processor during bus idle cycle, is:
- The status of S0 and S1 became low.
- The status of data, address and control bus in high impedance state.

56. What is wait state?

Answer: The WAIT state plays a significant role in preventing CPU speed incompatibilities.

Many times the processor is at ready state to accept data from a device or location, but there might be no input available. This can lead to wastage of CPU time.

So, in such cases when the CPU is ready for an input but there is no such valid data then the system gets into WAIT state. In this scenario the pin 35 (ready pin) is put into a low state.

As soon as there is some valid data for the 8085 the system comes off the WAIT state and the low state of the READY pin is withdrawn.

57. What is the difference between wait state and bus idle condition?

Answer: The difference is when the delay occurs: In a *"wait state"*, the delay occurs during a memory (or I/O) access cycle, giving the addressed device more time to respond. A *"bus idle"* state occurs between access cycles, and has no bearing on the addressed devices, other than giving them time to perform internal functions such as memory refresh.

58. How many instructions are available in 8085 instruction set?

Answer: 8085 processor has 246 instructions.

59. Which group of instruction affects the flags?

Answer: Arithmetic, Logical group of instruction affects the flags.

60. Name the arithmetic instructions that do not affect the flag?

Answer: Data transfer group is the arithmetic instruction which do not affect the flag.

61. List the instructions that affect only carry flag.

Answer: CMC, DAD, RAL, RAR, RLC, RRC and STC are the instructions that affect only carry flag.

62. What is DAA?

Answer: The working of DAA instruction depends on the contents of the AL register, CY, and AC flags. In effect, it adds 00H, 06H, 60H, or 66H to Accumulator to get the correct BCD answer in the Accumulator.

So, here is the illustration of the remedial actions against the previous example:

```
38 - - -> 0011 1000
  +45 - - -> 0100 0101
  ---------------------
  83    0111 1101
        -------  -------
         7       D
  0111 1101
    + 0110
  --------------
  1000 0011 - - - -> 83  (Decimal Sum)
```

As the summary of all related rules are listed below:

- If the LS hex digit in A is <= 9 and AC flag is 0, the LS hex digit value will not be altered.
- If the LS hex digit is >9, or if AC flag is set to 1, it adds 6 to the LS hex digit of A. If carry results, then it increments the MS hex

digit if this addition resulted in a carry to the MS digit position. In this process, the CY flag will be set to 1 if the MS hex digit was incremented from F to 0.

- If the MS hex digit is <= 9 and CY flag is 0, the MS hex digit will not be altered, and CY flag is reset to 0.
- If the MS hex digit is > 9, or if CY flag is set to 1, it adds 6 to the MS hex digit of A and sets CY flag to 1.

Note that for decimal subtraction DAA instruction cannot be used. Due to unavailability of decimal subtraction in Intel 8085 instruction set, a series of instructions are to be executed to perform decimal subtraction.

63. What is DAD instruction and what are the flags affected this instruction?

Answer: Add register pair to HL register. The 16-bit contents of the specified register pair are added to the contents of the HL register and the sum is stored in the HL register. The contents of the source register pair are not altered. If the result is larger than 16 bits, the CY flag is set. No other flags are affected.

64. List the various instructions that can be used to clear accumulator.

Answer: The instructions to reset the accumulator in 8085 are:

S.NO.	MNEMONICS	COMMENT
1	MVI A, 00	A <- 00
2	ANI 00	A AND 00
3	XRA A	A XOR A
4	SUB A	A <- A - A

65. What is difference between subtract and compare instruction?

Answer: Compare the numerical value of the destination with the source and set flags appropriately. This comparison is carried out in the form of a subtraction to determine which of the operands has a greater value. After a CMP instruction, OF, SF, ZF, and CF are set appropriately. For example, if the operands have equal values, then ZF if set.

The CMP instruction does not modify the destination field

SUB Subtracts the source value from the destination. Operation is almost identical to addition, except that the CF flag is used as a borrow in the case of the SBB (subtract with borrow) instruction.

66. What will be content of stack pointer after execution of PUSH and POP instructions?

Answer:

Program:

MEMORY ADDRESS	MNEMONICS
2000	LXI SP 3FFF
2003	PUSH H
2004	PUSH D
2005	POP H
2006	POP D
2007	HLT

Explanation:

Registers used are H, L, D, E

1. **LXI SP 3FFF:** Initialize SP by 3FFF.
2. **PUSH H:** Push the content of H and L register into the stack and decrements stack pointer by 2.
3. **PUSH D:** Push the content of D and E register into the stack and decrements stack pointer by 2.
4. **POP H:** Pop the upper two bytes from top of stack and place it in HL register pair and increment SP by 2.
5. **POP D:** Pop the upper two bytes from top of stack and place it in DE register pair and increment SP by 2.
6. **HLT:** Stops executing the program and halts any further execution.

67. What is the difference between call and jump instruction?

Answer: Jump instruction changes the value of program counter permanently, it can jump within the program memory (Program memory is starting from 2000H and end with 2100H). Whereas call instruction will change the value of program counter is short-term, it can jump throughout the memory i.e. 0000H to FFFFH.

68. Write a delay routine to produce a time delay of 0.5 msec in 8085 processor-based system whose clock source is 6 MHz quartz crystal.

Answer:

MVI D, N		; Load the count value, N in D-register.
Loop:	DCR D	; Decrement the count.
JNZ Loop		; If count is zero go to
RET		; Return to main program.

69. Write an ALP (Assembly language program) to addition of two 8-bit numbers.

Answer:

LXI H 4000H	: HL points 4000H
MOV A, M	: Get first operand
INX H	: HL points 4001H
ADD M	: Add second operand
INX H	: HL points 4002H
MOV M, A	: Store result at 4002H
HLT	: Terminate program execution

70. Write an ALP to subtraction of two 8-bit numbers.

Answer:

LXI H 4000H	: HL points 4000H
MOV A, M	: Get first operand
INX H	: HL points 4001H
SUB M	: Add second operand
INX H	: HL points 4002H
MOV M, A	: Store result at 4002H
HLT	: Terminate program execution

71. Write an ALP to multiplication of two 8-bit numbers.

Answer:

LXI H,8000H	; Load first operand address
MOV B, M	; Store first operand to B

INX H	; Increase HL pair
XRA A	; Clear accumulator
MOV C, A	; Store 00H at register C
LOOP ADD M	; Add memory element with Acc
JNC SKIP	; when Carry flag is 0, skip next task
INR C	; Increase C when carry is 1
SKIP DCR B	; Decrease B register
JNZ LOOP	; Jump to loop when Z flag is not 1
LXI H,8050H	; Load Destination address
MOV M, C	; Store C register content into memory
INX H	; Increase HL Pair
MOV M, A	; Store Acc content to memory
HLT	; Terminate the program

72. Write an ALP to division of two 8-bit numbers.
Answer:

LXI H, 5000	; Load first operand address
MOV B, M	; Get the dividend in B - reg.
MVI C, 00	; Clear C - reg for quotient
INX H	; Increment HL pair of registers
MOV A, M	; Get the divisor in A – reg
NEXT: CMP B	; compare A - reg with register B.
JC LOOP	; Jump on carry to LOOP
SUB B	; subtract A - reg from B - reg.
INR C	; Increment content of register C.
JMP NEXT	; Jump to NEXT
LOOP: STA 5002	; Store the remainder in Memory
MOV A, C	; Move Content of C - Reg to A - Reg
STA 5003	; Store the quotient in memory
HLT	; Terminate the program.

73. Write an ALP of addition of two 16-bit numbers.

Answer:

LDA 2050	; stores the value at 2050 in A (accumulator)
MOV B, A	; stores the value of A into B register
LDA 2052	; stores the value at 2052 in A
ADD B	; add the contents of B and A and store in A
STA 3050	; stores the result in memory location 3050
LDA 2051	; stores the value at 2051 in A
MOV B, A	; stores the value of A into B register
LDA 2053	; stores the value at 2053 in A
ADC B	; add the contents of B, A and carry from the lower bit addition and store in A
STA 3051	; stores the result in memory location 3051
HLT	; stops execution

74. Write an ALP to subtraction of two 16-bit numbers.

Answer:

LHLD 4000H	: Get first 16-bit number in HL
XCHG	: Save first 16-bit number in DE
LHLD 4002H	: Get second 16-bit number in HL
MOV A, E	: Get lower byte of the first number
SUB L	: Subtract lower byte of the second number
MOV L, A	: Store the result in L register
MOV A, D	: Get higher byte of the first number
SBB H	: Subtract higher byte of second number with borrow
MOV H, A	: Store 16-bit result in memory locations 4004H and 4005H.

SHLD 4004H	: Store 16-bit result in memory locations 4004H and 4005H.
HLT	: Terminate program execution.

75. Write an ALP to find the largest number in a given array.

Answer:

LXI H, 5000	;Set pointer for array
MOV B,M	;Load the Count
INX H	;Increment HL pair of registers
MOV A,M	;Copy content of memory from memory location pointed by HL pair of registers to Accumulator
DCR B	;Decrement B register
LOOP: INX H	;Increment HL pair of registers
AHEAD: DCR B	;Decrement B register
MOV A,M	;Set 1st element as largest data
DCR B	;Decrement the count
CMP M	;If A -reg> M go to AHEAD
JNC AHEAD	
MOV A,M	;Set the new value as largest
JNZ LOOP	;Repeat comparisons till count = 0
STA 6000	;Store the largest value at 6000
HLT	;Terminate Program

76. Write an ALP to find the smallest number in a given array.

Answer:

LXI H,5000	;Set pointer for array
MOV B,M	;Load the Count
INX H	
MOV A,M	;Set 1st element as largest data
DCR B	;Decrement the count
LOOP:	INX H
CMP M	;If A- reg< M go to AHEAD
JC AHEAD	

```
        MOV A,M              ;Set the new value as smallest
AHEAD:                       DCR B
        JNZ LOOP             ;Repeat comparisons till count = 0
        STA 6000             ;Store the largest value at 6000
        HLT
```

77. What is the address bus size in 8085?

Answer: 16-bit is the address bus size in 8085.

78. What is the maximum size of memory that can be interfaced with 8085?

Answer: 64Kb is the maximum size of memory that can be interfaced with 8085.

79. Calculate the number of address lines required to interface 8Kb of memory with 8085?

Answer: 8 Kb means 8X 1024 bytes= 2^{13}. This means 13 address lines are required.

80. What are the two types of Address Decoding?

Answer: Partial Decoding and absolute decoding are the two types of address decoding.

81. Which control signals generated while interfacing memory with 8085?

Answer: When 8085 wants to read and write into memory, it activates IO/M, RD, WR signals.

IO/M	RD	WR	Operation
0	0	1	8085 reads data from memory
0	1	0	8085 writes data into memory

By using these three signals two control signals Memory Read and Memory write signals are generated.

82. Calculate the number of Chip Select lines while interfacing 16 kb memory with 8085?

Answer: 16 kb means 16X1024 bytes which is equal to 2^{14} bytes. This means 14 address lines are required so number of chips select lines will be 16-14=2.

83. What is a programmable peripheral device?

Answer: If the functions performed by a peripheral device can be altered or changed by a program instruction then the peripheral device is called programmable device. Usually the programmable devices will have control registers. The device can be programmed by sending control word in the prescribed format to the control register.

84. What are the internal parts of 8255?

Answer: The internal parts of 8255 are port-A, port-B, port-C, and Control register. The ports can be programmed for either input or output function in different operating modes.

85. What are the operating modes of port -A 8255?

Answer: The port-A of 8255 can be programmed to work in anyone of the following operating modes as input or output port.
- **Mode-0:** Simple I/O port.
- **Mode-1:** Handshake I/O port
- **Mode-2:** Bidirectional I/O port

86. What is the purpose of 8255 PPI?

Answer: The 8255A is widely used as programmable, parallel I/O device. It can be programmed to transfer data under various conditions, from simple I/O to interrupt I/O.

87. Specify the bit of a control word for the 8255, which differentiates between the I/O mode and the BSR mode?

Answer: BSR mode D7=0, and I/O mode D5=1

88. Write down the output control signals used in 8255A PPI?

Answer:

The output control signals used in 8255A PPI are:
- **OBF:** Output Buffer Full
- **ACK:** Acknowledgement

- **INTR**: Interrupt request
- **INTE:** Interrupt Enable

89. What is the use of mode 2 in 8255A PPI?

Answer: The mode is used primarily in applications such as data transfer between two computers or floppy disk controller interface.

90. Write down the features of 8255A

Answer: The prominent features of 8255A are:

- It consists of three 8-bit IO ports i.e. PA, PB, and PC to enhance the flexibility of 8225.
- Address/data bus must be externally de-multiplexed.
- It is TTL compatible.
- It has improved DC driving capability.

91. What is handshake port?

Answer: In handshake port, signals are exchanged between device and port or between port and processor for checking/ information various condition of the device.

92. What are the features of Intel 8086?

Answer: Features of Intel 8086 are:

- Single +5V power supply
- Clock speed range of 5-10MHz
- Capable of executing about 0.33 Million instructions per second (MIPS)
- It is 16-bit processor having 16-bit ALU, 16-bit registers, internal data bus, and 16-bit external data bus resulting in faster processing.
- It uses two stages of pipelining, i.e. Fetch Stage and Execute Stage, which improves performance.
- Fetch stage can prefetch up to 6 bytes of instructions and stores them in the queue.
- It has 256 interrupts.

93. How many instructions can be executed per second in 8086/8088?

Answer: 1.5 Million instructions can be executed per second in 8086/8088.

94. What is Logical Address in 8086?

Answer: A memory address on the 8086 consist of two numbers usually written in hexadecimal and separated by colon representing segment and the offset. The combination of segment and offset is referred to as a logical address. In short **Logical Address = Segment : Offset**.

95. What is Effective Address in 8086?

Answer: In general, memory accesses take the form of the following example: MOV AX,[Base Reg. + Index Reg. + Constant]. This example copies a word size value into the register AX. Combined the three parameters in brackets determine what is called the **effective address**, which is simply the offset referenced by the instruction.

96. What is data and address size in 8086?

Answer: The 8086 can operate on either 8-bit or 16-bit data. The 8086 uses 20-bit address to access memory and 16-bit address to access I/O devices.

97. Write the flags in 8086?

Answer: The flags in 8086 are:
- Carry Flag **(CF)**
- Overflow Flag **(OF)**
- Parity Flag **(PF)**
- Trace Flag **(TF)**
- Auxiliary Flag **(AF)**
- Interrupt Flag **(IF)**
- Zero Flag **(ZF)**
- Direction Flag **(DF)**
- Sign Flag **(SF)**

98. Explain the function of M/IO in 8086?

Answer: The signal M/IO is used to differentiate the memory address and IO Address. When the processor is accessing memory locations M/IO is asserted high and when it is accessing I/O mapped devices, then it asserts low.

99. What is the function of BIU?

Answer: The BIU contains the circuit for physical calculations and a pre-coding instruction byte queue and it makes the bus signal available for external interfacing of devices.

100. What is the function of EU?

Answer: The EU contains the register set of 8086 except segment registers and IP. It has 16-bit ALU able to perform arithmetic and logic operations.

101. What is the size of instruction queue in 8086?

Answer: The queue length depends on the fetching speed and execution speed. Sometime queue may be restricted due to the space available on the CPU chip.

102. What are the Interrupts of 8086?

Answer: The interrupts of 8086 are INTR and NMI. The INTR is general maskable interrupt and NMI is non-maskable interrupt.

103. How clock signal is generated in 8086?

Answer: The 8086 does not have on-chip clock generation circuit. Hence the clock generator chip, 8284 is connected to the CLK pin of 8086. The clock signal supplied by 8284 is divided by three for internal use. The maximum internal clock frequency of 8086 is 5 MHz.

104. Explain CALL and RETURN?

Answer: CALL and RETURN calls 16-bit memory address of a subroutine. It is a 3-byte instruction that transfers the program sequence to a subroutine and saves the content of the **PC (Program Counter 16-bit register)**, the address of the next instruction on the stack decrements the stack pointer register by 2.

Jumps unconditionally to the memory location specified by the 2nd and 3rd bytes. This instruction is accompanied by a return instruction in the subroutine. The return instruction is used either to return a function value or to terminate the execution of a function.

The exit maybe from anywhere within the function body including loops or nested blocks, if the function returns a value, the *'return'* instruction in required.

105. What is the use of HLDA?

Answer: HLDA is the acknowledge signal for HOLD. It indicates whether the HOLD signal is received or not. HOLD and HLDA are used as the control signal for DMA operation.

106. What is the role of READY Pin?

Answer: READY is used by microprocessor to check whether a peripheral is ready to accept or transfer data. If READY pin is high, then the peripheral is ready for data transfer. If not, then microprocessor wait until READY goes high.

107. What is difference between instructions MUL and IMUL in 8086?

Answer:

The difference between instructions MUL and IMUL in 8086:

- **MUL:** Instruction is used for unsigned multiplication. This instruction multiplies bytes or words.
- **IMUL:** Integer Multiply Instruction is used for signed multiplication.

108. What are the flags available in 8086?

Answer:

The flags available in 8086:

- **Control Flags:** Direction, Interrupt, Trap
- **Condition Flags:** CY, AC, S, Z, P, OV

109. What are special registers available in 8086?

Answer:

Special registers available in 8086:

- Instruction Pointer
- Code Segment
- Data Segment
- Stack Segment

- Extra Segment

110. What is difference between SHIFT and ROTATE Instruction?

Answer:

The SHIFT and ROTATE Instruction:

- SHIFT and ROTATE commands are used to convert a number to another form where some bits are shifted or rotated.
- A rotate instruction is closed loop instruction. That is the data moved out at one end is put back in at the other end.
- The shift instruction loses the data that is moved out of the last bit locations.
- Basic difference between shift and rotate is shift command makes "fall of" bits at the end of the register. Where rotate command makes *"wrap around"* at the end of the register.

111. What is the use of test instruction in 8086?

Answer: Test instruction is same as the AND instruction except that it does not put result anywhere. Like the CMP instruction, it is used only to set flag register.

112. Which flags can be set or reset by the programmer and used to control the operation of the processor?

Answer: The flags can be set or reset by the programmer and used to control the operation of the processor:

- Trace Flag,
- Interrupt Flag,
- Direction Flag.

113. Which Segment is used to store interrupt and subroutine return address registers?

Answer: Stack Segment in segment register is used to store interrupt and subroutine return address registers.

114. What is the position of the stack pointer after the push instruction?

Answer: The address line is 02 less than the earlier value. Decrement by 2.

115. What is Internal Structure of 8086?

Answer: 8086 is having two different units i.e. Bus Interface Unit (BIU) and Execution Unit (EU), these two units work synchronously.

116. How many types memory can be divided in 8086?

Answer:

Memory is divided into two banks:
- even bank
- odd bank

117. Which Stack Is Used In 8086?

Answer: First In First Out (FIFO) stack is used in 8086. In this type of Stack, the first stored information is retrieved first.

118. Which microprocessor accepts the program written for 8086 without any changes?

Answer: 8088, 80186 microprocessor accepts the program written for 8086 without any changes

119. What does EU do?

Answer: Execution Unit (EU) Fetch the instruction from Queue (memory (6 byte) in BIU) and execute it.

120. Which Flags can be set or reset by the programmer and used to control the operation of the processor?

Answer: Trace Flag, Interrupt Flag, Direction Flag can be set or reset by the programmer and used to control the operation of the processor.

121. Which Segment is used to store interrupt and subroutine return address registers?

Answer: Stack Segment in segment register is used to store interrupt and subroutine return address registers

122. What is the difference between 8085 and 8086 in microprocessor?

Answer: 8086 has a special concept called as memory segmentation. It allows parallel processing, while 8085 does not.

123. What is idle state?

Answer: An idle state is a period of no bus activity that occurs because the pre fetch queue is full and the instruction current being executed does not require bus activity.

124. Define the function of BIU unit

Answer: BIU unit generates the 20-bit physical address for memory access. It fetches instruction from memory. It transfers data to and from the memory and I/O. It supports pipelining using the 6-byte instruction queue.

125. Define the function of instruction pointer (IP).

Function of instruction pointer (IP):

- It is a 16-bit register. It holds offset of the next instructions in the Code Segment.
- Address of the next instruction is calculated as CS x 10H + IP.
- IP is incremented after every instruction byte is fetched.
- IP gets a new value whenever a branch occurs.

126. Write down the function of execution unit?

Answer: The function of execution unit is:

- It fetches instructions from the Queue in BIU, decodes and executes them.
- It performs arithmetic, logic and internal data transfer operations within the microprocessor.
- It sends request signals to the BIU to access the external module.
- It operates with respect to T-stats (clock cycles) and does not depend upon which machine cycle is being performed by the BIU.

127. What is the need for memory segmentation?

Answer: The need for memory segmentation is:

- The Bus Interfacing Unit (BIU) contains four special purpose registers called as segment registers. These are Code Segment (CS) register, Stack Segment (SS) register, Extra Segment (ES) register and Data Segment (DS) register. All these are 16-bit registers.

- The number of address lines in 8086 is 20. So, the 8086 BIU will send out a 20-bit address in order to access one of the 1,048,576 or 1MB memory locations.
- But it is interesting to note that the 8086 does not work the whole 1MB memory at any given time. However, it works with only four 64 KB segments within the whole 1 MB memory.
- The four segment registers contain the upper 16 bits of the starting addresses of the four memory segments of 64 KB each with which the 8086 is working at that instant of time.
- Each segment is made up of memory contiguous memory locations It is independent, separately addressable unit.
- Segment registers are very useful for large programming tasks that require isolation of program code from the data code or isolation of module data from the stack information etc.
- Segmentation builds reloadable and re-entrant programs easily. In many cases the task of relocating a program simply requires moving the program code and then adjusting the code segment register to point to the base of the new code area.
- It allows to extend the address ability of a processor i.e. segmentation allows the use of 16-bit registers to give an addressing capability of 1 MB. Without segmentation, it would require 20-bit registers.

128. How to access 8-bit register in 8086 Microprocessor?

Answer: 8-bit register can be accessed by lower order and higher order of the individual 16-bit register like AL, BL, CL, DL.

129. How can you classify the 8086 register?

Answer:

8086 register can be classified as:
- Data registers
- Index registers
- Segment registers
- Pointer registers
- General registers

130. What are the rules available for memory segmentation?

Answer:

The rules available for memory segmentation are:

- The four segments can overlap for small programs. In a minimum system all four segments can start at the address 00000H.
- The segment can begin/start at any memory address which is divisible by 16.

131. Write the advantages of memory segmentation?

Answer:

The advantages of memory segmentation are:

- It allows the memory addressing capacity to be 1 Mbyte even though the address associated with individual instruction is only 16-bit.
- It allows instruction code, data, stack, and portion of program to be more than 64 KB long by using more than one code, data, stack segment, and extra.
- It facilitates use of separate memory areas for program, data and stack.
- It permits a program or its data to be put in different areas of memory, each time the program is executed i.e. program can be relocated which is very useful in multiprogramming.

132. Define LAHF and SAHF instructions in 8086.

Answer:

LAHF: load the 8086 equivalent flags into the AH register.

SAHF: store the AH register into the lower order byte of the flag register.

133. While handling the interrupt instructions in 8086. What are the internal operations?

Answer:

- SP = SP − 2, stack push flag register. Contents IF = 0, TF = 0
- SP = SP − 2, stack CS register. Contents, Address of interrupt pointer = interrupt type * 4, CS register contents second word of interrupt pointer.

- SP = SP – 2, Stack IP, IP first word interrupt pointer.

134. Write an ALP program in 8086 to add two 16-bit numbers.

Answer:

MOV AX, 1234H

MOV BX, 1234H

ADD AX, BX

HLT

135. Write about EVEN directive in 8086 ALP?

Answer: EVEN directive in 8086 ALP forces the address of the next byte to be even. 8086 words can be accessed in less time if they begin at even address.

136. Write an ALP program in 8086 to subtract two 16-bit numbers.

Answer:

MOV AX, 1234H

MOV BX, 1234H

SUB AX, BX

HLT

137. Define multiprocessor systems.

Answer: Microprocessor system contains two or more components that can execute instructions independently, then the system is called multiprocessor.

138. List the advantages of multiprocessor systems.

Answer:

The advantages of multiprocessor systems are:

- Improves cost/performance ratio of the system.
- Several processors may be combined to fit the needs of an application while avoiding the expense of the unneeded capabilities of a centralized system.
- Tasks are divided among the modules. If failure occurs, it is easier and cheaper to find and replace the malfunctioning processor then replacing the failing part of complex processor.

139. Describe the move instruction in 8086 microprocessor.

Answer: The MOV instructions are of various types as discussed below:

MOV reg, data

This instruction moves immediate 8 bit or 16-bit data to the specified register. Here register is used either 8-bit register or 16-bit register.

Example:

8-bit register: AL, BL, CL, DL.

16-bit register: AX, BX, CX, DX, SI and DI.

MOV AL, 23H

MOV BX, 1234H

140. Describe the multiplication instruction in 8086 microprocessor.

Answer: MUL mem/reg: This instruction multiplies content of AL or AX register with content of specified memory or register.

141. Describe the division instruction in 8086 microprocessor.

Answer: DIV mem/reg: This instruction divides the content of AL or AX register with content of specified memory or register.

142. Give short notes on SUB instruction.

Answer: SUB instruction will perform subtraction of two 8-bit data or 16-bit data, which is stored in memory, in register or immediate data.

Example:

SUB ac, data

This instruction subtracts the immediate data to AL or AX register and store the result in accumulator.

143. Differentiate between call and jump instruction in 8086?

Answer:

- A JMP instruction permanently changes the program counter. A CALL instruction leaves information on the stack so that the original program execution sequence can be resumed.
- CALL is an instruction that transfers the program control to a sub routine with the intention of coming back to the main program.

- Thus, in CALL 8086 saves the address of the next instruction into the stack before branching to the sub routine.
- At the end of the sub routine, control transfers back to the main program using the return address of the stack. There are two types of CALL: Near CALL and Far CALL.

144. What is the function of call instruction while calling inter segment(near)?

Answer:
1. The new subroutine called must be in the same segment (hence intra segment).
2. The CALL address can be specified directly in the instruction or indirectly through registers or memory locations.
3. The following sequence is executed for a NEAR CALL:
o 8086 will PUSH current IP into the stack.
o Decrement SP by 2.
o New value loaded into IP.
o Control transferred to a sub routine within the same segment.

145. What is the function of call instruction while calling inter segment (Far)?

Answer: The function of call instruction while calling inter segment are:
1. The new sub routine called is in another segment (hence inter segment).
2. Here, CS and IP both get new values.
3. The CALL address can be specified directly or through registers or memory locations.
4. The following sequence is executed for a FAR CALL:
o PUSH CS into the stack.
o Decrement SP by 2.
o PUSH IP into the stack.
o Decrement SP by 2.
o Load CS with new segment address.
o Load IP with new offset address.

o Control transferred to a sub routine in the new segment.

146. What is the function of jump instruction while calling intra segment (Near)?

Answer:

The jump address is specified in two ways:

- **INTRA segment direct jump**: The new branch location is specified directly in the instruction.
- The new address is calculated by adding the 8 or 16-bit displacement to the IP. The CS does not change.
- A positive displacement means that the jump is ahead in the program. A negative displacement means that the jump is behind in the program. It is also called **relative jump**.
- **INTRA segment indirect jump**: The new branch address is specified indirectly through a register or a memory location. The value in the IP is replaced with the new value. The CS does not change.

147. What is the function of jump instruction while calling inter segment (Far)?

Answer: The jump address is specified in two ways:

- **INTER segment direct jump**: The new branch location is specified directly in the instruction. Both CS and IP get new values as this is an inter segment jump.
- **INTER segment indirect jump**: The new branch location is specified indirectly through a register or a memory location. Both CS and IP get new values as this is an inter segment jump.

148. What are the tasks performed by microprocessor 8086 when an interrupt encounter?

Answer: The task performed by microprocessor when an interrupt encounter are:

- The value of flag register is pushed into the stack. It means that first the value of SP (Stack Pointer) is decremented by 2 then the value of flag register is pushed to the memory address of stack segment.
- The value of starting memory address of Code Segment (CS) is pushed into the stack.

- The value of Instruction Pointer (IP) is pushed into the stack.
- IP is loaded from word location (Interrupt type) * 04.
- CS is loaded from the next word location.
- Interrupt and Trap flag are reset to 0.

149. Discuss hardware interrupt in 8086.

Answer: Hardware interrupts are caused by any peripheral device by sending a signal through a specified pin to the microprocessor. There are two hardware interrupts in 8086 microprocessors:

a. NMI (Non Maskable Interrupt): It is a single pin non maskable hardware interrupt which cannot be disabled. It is the highest priority interrupt in 8086 microprocessors. After its execution, this interrupt generates a TYPE 2 interrupt. IP is loaded from word location 00008 H and CS is loaded from the word location 0000A H.

b. **INTR (Interrupt Request):** It provides a single interrupt request and is activated by I/O port. This interrupt can be masked or delayed. It is a level triggered interrupt. It can receive any interrupt type, so the value of IP and CS will change on the interrupt type received.

150. Discuss software interrupt in 8086.

Answer: Software interrupt are instructions within the program to generate interrupts. There are 256 software interrupts in 8086 microprocessor. The instructions are of the format INT type where type ranges from 00 to FF. The starting address ranges from 00000 H to 003FF H. These are two-byte instructions. IP is loaded from type * 04 H and CS is loaded from the next address give by (type * 04) + 02 H.

INT nn is invoked software (sequence of code)

- *TYPE 0* corresponds to division by zero (0).
- *TYPE 1* is used for single step execution for debugging of program.
- *TYPE 2* represents NMI and is used in power failure conditions.
- *TYPE 3* represents a break-point interrupt.
- *TYPE 4* is the overflow interrupt.

Example:
- DOS INT 21H, BIOS INT 10H.
- INT 00 (divide error)

- INT 01 (single step)
- INT 03 (breakpoint)
- INT 04 (signed number overflow)

151. Define interrupt priority.

Answer: If two or more interrupts occur at the same time then the highest priority interrupt will be serviced first, and then the next highest priority interrupt will be serviced.

Example: If INTR input is enabled, and 8086 receives an INTR signal during the execution of a divide instruction, and the divide operation produces a divide by zero interrupt. Since the internal interrupts- such as divide error, INT, and INTO have higher priority than INTR the 8086 will perform a divide error interrupt response first. The interrupt that has a lower address, has a higher priority.

152. What are the bit manipulation instructions are there in 8086 microprocessor?

Answer: Bit manipulation or logical instructions in microprocessor 8086 are:
- NOT
- AND
- OR
- XOR
- TEST

153. Describe AND instruction in 8086.

Answer: AND instruction logically ANDs each bit of the source byte or word with the corresponding bit the destination and stores result in the destination. The source may be immediate number, a register or a memory location. The destination may be a register or a memory location.

Syntax:

    AND destination, source

Example: AND BL, CL

        AND AX, BX

154. Describe about OR instruction in 8086.

Answer: OR instruction logically ORs each bit of the source byte or word with the corresponding bit the destination and stores result in the destination. The source may be immediate number, a register or a memory location. The destination may be a register or a memory location.

Syntax:

OR destination, source

Example: OR BL, CL

OR AX, BX

155. Describe XOR instruction in 8086.

Answer: XOR instruction logically XORs each bit of the source byte or word with the corresponding bit the destination and stores result in the destination. The source may be immediate number, a register or a memory location. The destination may be a register or a memory location.

Syntax:

XOR destination, source

Example: XOR BL, CL

XOR AX, BX

156. Write an ALP to add two 8-bit numbers with 8086.

Answer:

MOV AL,12H

MOV BL, 12H

ADD AL, BL

MOV SI, 1200H

MOV [SI], AL

HLT

157. Write an ALP to subtract two 8-bit numbers with 8086.

Answer:

MOV AL, 12H

MOV BL, 12H

SUB AL, BL

MOV SI, 1200H
MOV [SI], AL
HLT

158. Write an ALP to add two 16-bit numbers with 8086.

Answer:

MOV AX, 1234H
MOV BX, 1234H
ADD AX, BX
MOV SI, 1200H
MOV [SI], AX
HLT

159. Write an ALP to subtract two 16-bit numbers with 8086.

Answer:

MOV AX, 1234H
MOV BX, 1234H
SUB AX, BX
MOV SI, 1200H
MOV [SI], AX
HLT

160. Write an ALP to multiply the two 8-bit numbers with 8086.

Answer:

MOV SI, 2000H
MOV DI, 2002H
MOV AL, [SI]
INC SI
MOV BL, [SI]
MUL BL
MOV [DI], AX
HLT

161. Write an ALP to multiply the two 16-bit numbers with 8086.

Answer:
MOV AX, [3000]
MOV BX, [3002]
MUL BX
MOV [3004], AX
MOV AX, DX
MOV [3006], AX
HLT

162. Write an ALP to divide the two 8-bit numbers with 8086.

Answer:
MOV SI, 2000H
MOV DI, 2002H
MOV AL, [SI]
INC SI
MOV BL, [SI]
DIV BL
MOV [DI], AX
HLT

163. Write an ALP to divide the two 16-bit numbers with 8086.

Answer:
MOV AX, [3000]
MOV BX, [3002]
DIV BX
MOV [3004], AX
MOV AX, DX
MOV [3006], AX
HLT

164. Write an ALP to find the square root of given number with 8086.

Answer:

    MOV AX, [2000]
    MOV CX, 0000H
    MOV BX, FFFFH
    LOOP ADD BX, 02
    INC CX
    SUB AX, BX
    JNZ LOOP
    MOV [2500], CX
    HLT

165. Write an ALP to add two 8-bit BCD numbers with 8086.

Answer:

    MOV AL, [2000]
    MOV BL, [2001]
    ADD AL, BL
    DAA
    MOV [2002], AL
    MOV AL, 00H
    ADC AL, AL
    MOV [2003], AL
    HLT

166. Write an ALP to find sum of digits of 8-bit numbers with 8086.

Answer:

    MOV AL, [2000]
    MOV AH, AL
    MOV CX, 0004H
    AND AL, 0FH
    ROL AH, CX

AND AH, 0FH
ADD AL,AH
MOV [2051], AL
HLT

167. Explain Pins 1 to 8 of 8051 Micro controller.

Answer: Pins 1 to 8 of 8051 Microcontroller are known as Port 1. This port doesn't serve any other functions. It is internally pulled up, bi-directional I/O port.

168. Explain Pins 10 to 17 of 8051 Micro controller.

Answer: Pins 10 to 17 of 8051 Micro controller are known as Port 3. This port serves some functions like interrupts, timer input, control signals, serial communication signals RxD and TxD, etc.

169. Explain Pins 21 to 28 of 8051 Micro controller.

Answer: Pins 21 to 28 of 8051 Micro controller are known as Port 2. It serves as I/O port. Higher order address bus signals are also multiplexed using this port.

170. Explain Pins 32 to 39 of 8051 Micro controller.

Answer: Pins 32 to 39 of 8051 Micro controller are known as Port 0. It serves as I/O port. Lower order address and data bus signals are multiplexed using this port.

171. Explain Pin 18 and 19 of 8051 Micro controller.

Answer: X2 and X1 pins are input output pins for the oscillator. These pins are used to connect an internal oscillator to the microcontroller.

172. What is the reset pin of 8051 Micro controller?

Answer: Pin 9 is the Reset Input Pin. It is an active HIGH Pin i.e. if the RST Pin is HIGH for a minimum of two machine cycles, the microcontroller will be reset. During this time, the oscillator must be running.

173. Explain the ALE/PROG pin of 8051 Micro controller.

Answer: Pin 30 is the Address Latch Enable, using this pin, external address can be separated from data (as they are multiplexed by 8051).

174. Intel 8051 follows which architecture?

Answer: Intel 8051 follows Harvard Architecture.

175. What is the difference between Harvard Architecture and von Neumann Architecture?

Answer: The most obvious characteristic of the Harvard Architecture is that it has physically separate signals and storage for code and data memory. Von Neumann machines have shared signals and memory for code and data. Thus, the program can be easily modified by itself since it is stored in read-write memory.

176. 8051 was developed by using which technology?

Answer: Intel's original MCS-51 family was developed by using NMOS technology, but later versions, identified by a letter C in their name (e.g., 80C51) used CMOS technology and consume less power than their NMOS predecessors.

177. Why 8051 is called 8-bit microcontroller?

Answer: The Intel 8051 is an 8-bit microcontroller which means that most available operations are limited to 8 bits.

178. What is the width of data bus?

Answer: The width of data bus is 8-bit.

179. What is the width of address bus?

Answer: The width of address bus is 16-bit.

180. List the features of 8051 micro controller.

Answer: The features of 8051 micro controller are:

40 Pin IC., 128 bytes of RAM, 4K ROM, 2 Timers (Timer 0 and Timer 1), 32 Input/Output pins, 1 serial port, 6 Interrupts (Including Reset).

181. On-chip RAM is also called _____ memory?

Answer: Direct memory

182. What location code memory space and data memory space begins?

Answer: At location 0x00 for internal or external memory

183. How Much on chip RAM is available?

Answer: 128 bytes of RAM (from 0x00 to 0x7F) and can be used to store data.

184. Internal RAM is located from address 0x00 to ___?

Answer: Internal RAM in 8051 is located from address 0 to address 0xFF. IRAM from 0x00 to 0x7F can be accessed directly. IRAM from 0x80 to 0xFF must be accessed indirectly.

185. With 12 MHz clock frequency how many instructions (of 1 machine cycle and 2 machine cycle) it can execute per second?

Answer: A cycle is 12 pulses of the crystal. That is to say, if an instruction takes one machine cycle to execute, it will take 12 pulses of the crystal to execute. Since we know the crystal is pulsing 11,059,000 times per second and that one machine cycle is 12 pulses, we can calculate how many instructions cycle the 8051 can execute per second

186. List the addressing Modes in MCS-51

Answer: The addressing Modes in MCS-51 are:

Direct Addressing, Register Addressing, Register Indirect Addressing, Implicit Addressing, Immediate Addressing, Index Addressing

187. How much total external data memory that can be interfaced to the 8051?

Answer: 64K data memory

188. What is Special Function Registers (SFR)?

Answer: The memory addresses from 80H to 0FFH are called SFR. These are 128 bytes registers specially designed for interrupts and few other operations.

189. What are the four distinct types of memory in 8051?

Answer: The four distinct types of memory in 8051 are:

Internal RAM, Special function registers, Program memory, External data memory

190. Can single bit of a port be accessed in 8051?

Answer: Yes, 8051 has the capability of accessing only single bit of a port, Here, only single bit is accessed, and rest is unaltered.

191. Other than SETB, CLR are there any single bit instructions?

Answer: There are in total 6 single-bit instructions, CPL bit: complement the bit (bit= NOT bit), JB bit, target: Jump to target if bit equal to 1.

192. What are the types of interrupts in 8051?

Answer: Types of interrupts in 8051 are:

External interrupt 0 (IE0) has highest priority among interrupts, Timer interrupt 0 (TF0), External interrupt 1 (IE1), Timer interrupt 1 (TF1) has lowest priority among other interrupts, Serial port Interrupt, Reset.

193. Tell the addresses which are bit addressable?

Answer: The bit addressable memory in 8051 is compose from 210 bits: bit address space: 20H – 2FH bytes RAM = 00H – 7FH bits address, SFR registers.

194. What is lst file?

Answer: Lst file is also called as list file, It lists the opcodes, addresses and errors detected by the assembler, List file is produced only when indicated by the user.

195. Explain DB.

Answer: DB is called as define byte used as a directive in the assembler, it is used to define the 8-bit data in binary, hexadecimal or decimal formats.

196. What is EQU?

Answer: EQU is the equate assembler directive used to define a constant without occupying a memory location, It associates a constant value with data label.

197. How are labels named in assembly language?

Answer: Label name should be unique and must contain alphabetic letters in both uppercase and lowercase, 1st letter should always be an alphabetic letter.

198. Which bit of the flag register is set when output overflows to the sign bit?

Answer: The 2nd bit of the flag register is set when output flows to the sign bit. This flag is also called as the overflow flag. Here the output of the signed number operation is too large to be accommodated in 7 bits.

199. What are issues related to stack and bank 1.

Answer: Bank 1 uses the same RAM space as the stack, Stack pointer is incremented or decremented according to the push or pop instruction.

200. Explain JNC.

Answer: JNC is an instruction used to jump if no carry occurs after an arithmetic operation. It is called as jump if no carry (conditional jump instruction).

201. Caz port 0 be used as input output port in 8051?

Answer: Yes, port 0 can be used as input/output port. Port 0 is an open drain unlike ports 2, 3, 4.

202. Which two ports can be combined to form the 16-bit address for external memory access in 8051?

Answer: Port0 and port2 together form the 16-bit address for external memory, Port0 uses pins 32 to 39 of 8051 to give the lower address bits (AD0-AD7), Port2 uses pins 21 to 28 of 8051 to give the higher address bits (A8-A15).

203. How many timers and counters are there in 8051 microcontroller?

Answer: There are two 16-bit timers and counters in 8051 microcontroller: timer 0 and timer 1. Both timers consist of 16-bit register in which the lower byte is stored in TL and the higher byte is stored in TH.

204. Explain the special function registers of 8051 microcontroller?

Answer: Counters and Timers in 8051 microcontroller has two special function registers: Timer Mode Register (TMOD) and Timer Control Register (TCON), which are used for activating and configuring timers and counters.

205. What is the working of Timer Mode Control (TMOD) of 8051 micro-controller?

Answer: TMOD is an 8-bit register used for selecting timer or counter and mode of timers. Lower 4-bits are used for control operation of timer 0 or counter0 and remaining 4-bits are used for control operation of timer1 or counter1.

206. What is the function of gate in 8051 micro-controller?

Answer: Gate: If the gate bit is '0', then we can start and stop the "software" timer in the same way. If the gate is set to '1', then we can perform hardware timer.

207. Explain the different modes of timers in 8051 micro-controller?

Answer: The different modes of timers in 8051 microcontroller:

- **Mode 0:** This is a 13-bit mode that means the timer operation completes with *"8192"* pulses.
- **Mode 1:** This is a 16-bit mode, which means the timer operation completes with maximum clock pulses that *"65535"*.
- **Mode 2:** This mode is an 8-bit auto reload mode, which means the timer operation completes with only *"256"* clock pulses.
- **Mode 3:** This mode is a split-timer mode, which means the loading values in T0 and automatically starts the T1.

208. What is Timer Control Register (TCON) of 8051 micro-controller?

Answer: TCON is a register used to control operations of counter and timers in microcontrollers. It is an 8-bit register wherein four upper bits are responsible for timers and counters and lower bits are responsible for interrupts.

209. Explain the TF1 register in 8051 micro-controller?

Answer: TF1 stands for *'timer1'* flag bit. Whenever calculating the time-delay in timer1, the TH1 and TL1 reaches to the maximum value that is *"FFFF"* automatically.

210. Explain the TR1 register in 8051 micro-controller?

Answer: The TR1 stands for timer1 start or stop bit. This timer starting can be through software instruction or through hardware method.

211. What is the Procedure to Calculate the Delay Program in 8051 micro-controller?

Answer: First load the TMOD register value for *'Timer0'* and *'Timer1'* in different modes. Whenever we operate the timer in mode 1, timer takes the maximum pulses of 65535. Start the timer1 "TR1=1;" Monitor the flag bit "while(TF1==1)", Clear the flag bit "TF1=0", Cleat the timer "TR1=0".

212. What is the difference between serial and parallel communication?

Answer: Parallel communication is fast, but it is not applicable for long distances (for printers). Moreover, it is also expensive. Serial is not much fast as parallel communication, but it can deal with transmission of data over longer distances (for telephone line, ADC, DAC).

213. Explain serial communication of 8051 micro controller.

Answer: The register SBUF is used to hold the data. The special function register SBUF has two registers. One is, write-only and is used to hold data to be transmitted out of the 8051 via TXD.

214. Explain the methods of serial communication in 8051 micro controller.

Answer: The methods of serial communication in 8051 micro controller are:

- **Synchronous Communication:** Transfer of bulk data in framed structure at a time
- **Asynchronous Communication:** Transfer of a byte data in framed structure at a time

215. Explain Data transmission rate in 8051 micro controller.

Answer: Data transmission rate is measured in bits per second (bps). In binary system it is also called as baud rate (number of signal changes per second). Standard baud rates supported are 1200, 2400, 4800, 19200, 38400, 57600, and 115200.

216. Explain the difference between USART and UART.

Answer:

- **USART** stands for Universal Synchronous/Asynchronous Receiver-Transmitter. USART uses external clock so it needs separate line to carry the clock signal.
- **UART** stands for Universal Asynchronous Receiver-Transmitter. A UART generates its internal data clock to the microcontroller. It synchronizes that clock with the data stream by using the start bit transition.

217. At which pin we can attach the LED using 8051 micro controller?

Answer: At any pin of the port.

218. Which pin does not consist the internal pull up in 8051 microcontroller?

Answer: Port 0

219. Which configuration is preferred to apply LED with 8051 microcontroller?

Answer: Common anode.

220. How many LED's can be attached directly with 8051 microcontroller.

Answer: Thirty-Two

221. Which port can be used as input or output port in 8051 micro controller.

Answer: All ports Po,P1,P2 and P3

222. Which pin does not consist the internal pull up in 8051 micro controller.

Answer: Port 0

223. Which configuration is preferred to apply LED and switch with 8051 micro controller.

Answer: Common anode and pull down.

224. How many switches can be attached directly with 8051 micro

controller.

Answer: Thirty-Two

225. What are the uses of Seven segment displays?

Answer: Seven segment displays are used to indicate numerical information. Seven segments display can display digits from 0 to 9 and even we can display few characters like A, b, C, H, E, e, F, etc.

226. What is the internal structure of seven segment displays?

Answer: Seven segment displays internally consist of 8 LEDs. In these LEDs, 7 LEDs are used to indicate the digits 0 to 9 and single LED is used for indicating decimal point. Generally, seven segments are two types, one is common cathode and the other is common anode.

227. Maximum how many seven segment displays can be connected to 8051.

Answer: Twenty-Four

228. How to drive more than one seven segments using 8051 micro controller?

Answer: Using the transistor one can drive more than one seven segments using 8051 micro controller.

229. Discuss details of 16x2 LCD

Answer: A 16 2 LCD module is a very common type of LCD module which is used in 8051 based embedded projects. It has 16 rows and 2 columns [5 7] or [5 8] LCD dot matrices.

230. How many control lines are there in 16x2 LCD?

Answer: Three

231. How many data lines are there in LCD?

Answer: 8 bits

232. How to connect LCD with 8051 micro controller?

Answer: Control lines to be connected to 3 pins of 8051 and data lines can be connected either by using 8 pins, 4 pins or 1 pin also.

233. How to glow back light of LCD.

Answer: Pin number 15 and 16 to be connected to Vcc and ground.

234. How to control the contrast of LCD

Answer: LCD contrast is controlled by applying the variable power supply to pin 3.

235. Give one advantage on LCD over seven segments

Answer: LCD can display alphanumeric character while seven segments is suitable for numeric character display.

236. Give one disadvantage on LCD over seven segments

Answer: LCD requires backlight to be viewed in dark while seven segments are made of LEDs which are light emitting diodes and thus can be easily viewed in dark.

237. What is the function of RS pin in LCD?

Answer: RS is Register Select pin, LCD has two types of registers, data and command register. This pin is used to select one register out of the two.

238. How to select data or command register in LCD

Answer: To select command register RS pin is to be made 0 and for selecting data register RS pin is to be made 1.

239. What is the function of Enable pin?

Answer: Whenever any data or command is sent to LCD, HIGH to LOW enable signal is required by the LCD to latch the information present at it pin.

240. What is the length of LCD data bus?

Answer: LCD has 8-bit data bus from D0 to D7.

241. What is the significance of 16x2 in LCD specification?

Answer: 16x2 signifies that LCD has 16 rows and 2 columns.

242. How to select character position on 16x2 LCD?

Answer: Address of first row on LCD start from 0x80 and goes through 0x8F. Address of second row on LCD start from 0xC0 and goes through 0xCF.

243. How to program LCD in 4-bit mode?

Answer: In order to program LCD in 4-bit mode the upper 4-bit of LCD data bus is connected to microcontroller and LCD is initialized in 4 -bit mode by the hex command 0x20.

244. What is the length of LCD data bus?

Answer: LCD has 8-bit data bus from D0 to D7.

245. How to select character position on 16x2 LCD

Answer: Address of first row on LCD start from 0x80 and goes through 0x8F. Address of second row on LCD start from 0xC0 and goes through 0xCF.

246. Explain the keypad matrix.

Answer: A keypad is a set of buttons arranged in a block or *"pad"* which usually bear digits, symbols and a complete set of alphabetical letters. It mostly contains numbers then it can also be called a numeric keypad.

247. How to interface the keypad matrix with 8051 Micro controller?

Answer: A 4x4 keypad matrix can be interfaced with the help of any port of 8015 Micro controller.

248. How does keypad matrix scan the row and column?

Answer:

1. First connect all the Rows to Logic level 0 and all the columns to Logic level 1.

2. Whenever we press a button, column and row, corresponding to that button gets shorted and makes the corresponding column to logic level 0. Because that column becomes connected (shorted) to the row, which is at Logic level 0. So, we can get the column no.

249. Which port can be used for keypad interfacing?

Answer: Any port except port 0. For port 0 the external pull, up need to be interfaced.

250. Define the different pin architecture available with PIC16F87x.

Answer: 28, 40 and 44 pins.

251. Does the Pin out is compatible to other 28-pin or 40/44-pin PIC16CXXX and PIC16FXXX microcontrollers?

Answer: Yes

252. What are the other devices similar to PIC16F877?

Answer: PIC16F873A   PIC16F876A   PIC16F874A   and PIC16F877A.

253. Give some details of PIC16F877.

Answer: PIC 16F877 is one of the most advanced microcontroller from Microchip. This controller is widely used for experimental and modern applications because of its low price, wide range of applications, high quality, and ease of availability.

254. What are the General Features of PIC16F877?

Answer: High performance RISC CPU, ONLY 35 simple word instructions, All single cycle instructions except for program branches which are two cycles, Operating speed: clock input (200MHz), instruction cycle (200nS), Up to 368 8bit of RAM (data memory), 256 8 of EEPROM (data memory), 8k 14 of flash memory, Pin out compatible to PIC 16C74B, PIC 16C76, PIC 16C77. Eight level deep hardware stack.

255. What is the operating speed of the PIC16F877?

Answer: 20 MHz clock input

256. Does PIC support In-Circuit Serial Programming?

Answer: Yes

257. Give the special features in PIC16f877 compared to AVR family microcontroller.

Answer: Watchdog Timer (WDT) with its own on-chip RC oscillator for reliable operation

In-Circuit Debug (ICD) via two pins, Power saving Sleep mode and Low cost

258. How much flash memory is available with PIC16F873A and PIC16F877A?

Answer: 4K and 8K.

Exercise 107

259. How much data memory is available with PIC16F873A and PIC16F877A?

Answer: 128 bytes and 256 bytes.

260. Give the details of Program Counter (PC) in PIC.

Answer: The Program Counter (PC) is 13 bits wide. The low byte comes from the PCL register which is a readable and writable register. The upper bits (PC<12:8>) are not readable but are indirectly writable through the PCLATH register.

261. What is Program Memory Paging, explain it?

Answer: All PIC16F87XA devices are capable of addressing a continuous 8K word block of program memory. The CALL and GOTO instructions provide only 11 bits of address to allow branching within any 2K program memory page. When doing a CALL or GOTO instruction, the upper 2 bits of the address are provided by PCLATH.

262. What is stack in PIC microcontroller. Give the details of stack.

Answer: The PIC16F87XA family has an 8-level deep x 13-bit wide hardware stack. The stack space is not a part of either program or data space and the stack pointer is not readable or writable. The PC is PUSHed onto the stack when a CALL instruction is executed, or an interrupt causes a branch. The stack is POP'ed in the event of a RETURN, RETLW or a RETFIE instruction execution. PCLATH is not affected by a PUSH or POP operation.

263. Explain the difference between Indirect Addressing, INDF and FSR Registers in PIC.

Answer: The INDF register is not a physical register. Addressing the INDF register will cause indirect addressing. Indirect addressing is possible by using the INDF register. Any instruction using the INDF register accesses the register pointed to by the File Select Register, FSR. Reading the INDF register itself, indirectly (FSR = 0) will read 00h. Writing to the INDF register indirectly results in a no operation (although status bits may be affected).

264. Give the step for Writing to Data EEPROM Memory of PIC.

Answer: To write an EEPROM data location, the user must first write the address to the EEADR register and the data to the EEDATA register. Then the user must follow a specific write sequence to initiate the write for each byte.

265. What are the steps for Reading Flash Program Memory?

Answer: To read a program memory location, the user must write two bytes of the address to the EEADR and EEADRH registers, set the EEPGD control bit (EECON1<7>) and then set control bit RD (EECON1<0>). Once the read control bit is set, the program memory flash controller will use the next two instruction cycles to read the data.

266. Which ports are available with PIC16F873A and PIC16F877A?

Answer: Ports A, B, C of 8 bits and Ports A, B, C, D, E of 8 bits.

267. Define PORTA and the TRISA Register of PIC.

Answer: PORTA is a 6-bit wide, bidirectional port. The corresponding data direction register is TRISA. Setting a TRISA bit (= 1) will make the corresponding PORTA pin an input (i.e., put the corresponding output driver in a High-Impedance mode).

268. Define PORTC and the TRISC Register of PIC.

Answer: PORTC is an 8-bit wide, bidirectional port. The corresponding data direction register is TRISC. Setting a TRISC bit (= 1) will make the corresponding PORTC pin an input (i.e., put the corresponding output driver in a High-Impedance mode).

269. How many input output ports are there in PIC 16F877?

Answer: PIC 16F877 series normally has five input/output ports.

270. What are the major functions of the ports of PIC16F877.

Answer: Most of these port pins are multiplexed for handling alternate function for peripheral features on the devices. All ports in a PIC chip are bi-directional.

271. Explain the working of PORT B and the TRIS B Registers.

Answer: PORT B is also an 8-bit bi-directional PORT. Its direction controlled and maintained by TRIS B data direction register. Setting the TRIS B into logic '1' makes the corresponding "PORT B" pin as an input. Clearing the TRIS B bit make PORT B as an output.

272. Explain the working of TRIS register.

Answer: TRIS register controls the Parallel Slave PORT operation. PORT pins are multiplexed with analog inputs. When selected for analog input, these pins will read as '0's.

273. Explain the working of PORT D and TRIS D Registers.

Answer: PORT D is an 8-bit PORT with bi-directional nature. This port also with Schmitt Trigger input buffers, each pin in this PORT D individually configurable as either input or output.

274. Explain the working of PORT E and TRIS E Registers.

Answer: PORT E has only three pins (RE0/RD/AN5, RE1/WR/AN6 and RE2/CS/AN7) which are individually configurable as inputs or outputs. These pins controllable by using its corresponding data direction register "TRIS E".

275. Define the timer 01 used in PIC16F877.

Answer: It is of 16-bit timer/counter with prescaler, can be incremented during Sleep via external crystal/clock.

276. Why Timer is used for delay modules?

Answer: A Delay macro is called a *"dump"* delay, because during the execution of Delay function the MCU sits dump by just creating a delay. During this process the MCU cannot listen to its ADC values or read anything from its Registers.

277. What are the short comings in delay macros?

Answer: The value of delay must be a constant for delay macros; it cannot be changed during program execution. Hence it remains is programmer defined. The delay will not be accurate as compared to using Timers.

278. Explain about the PIC microcontroller Timers.

Answer: The PIC16F877A PIC MCU has three Timer Modules. They are names as Timer0, Timer1 and Timer2. The Timer 0 and Timer 2 are 8-bit Timers and Timer 1 is a 16-bit Timer.

279. What is the difference between 8-bit and 16-bit timer?

Answer: 16-bit Timer has much better Resolution that the 8-bit Timer.

280. What is prescaler. Explain it?

Answer: Prescaler is a name for the part of a microcontroller which divides oscillator clock before it will reach logic that increases timer status.

281. What is the range of prescaler?

Answer: The range of the prescaler id is from 1 to 256 and the value of the Prescaler can be set by using the OPTION Register.

282. How the timer count from beginning to final value.

Answer: The Timer start incrementing once set and overflow after reaching a value of 256, to enable the Timer interrupt during this point the register TMR0IE has to be set high. Since Timer 0 itself is a peripheral we have to enable the Peripheral Interrupt by making PEIE=1.

283. What do you mean by interrupt service routine?

Answer: The interrupt service routine is an Interrupt that will be called each time the Timer0 is overflows.

284. How to read data from EEPROM Memory from PIC microcontroller.

Answer: To read a data memory location, the user must write the address to the EEADR register, clear the EEPGD control bit (EECON1) and then set control bit RD (EECON1<0>). The data is available in the very next cycle in the EEDATA register; therefore, it can be read in the next instruction. EEDATA will hold this value until another read or until it is written to by the user (during a write operation).

285. What are modes of USART PIC16F877.

Answer: There are two different modes namely the 8-bit and 9-bit mode. Asynchronous mode is with 8-bit communication system.

286. What is working of the asynchronous mode of communication in PIC16F877.

Answer: Asynchronous doesn't need to send clock signal along with the data signals. UART uses two data lines for sending (Tx) and receiving (Rx) data. The ground of both devices should also be made common.

287. How to convert RS232 to USB converter.

Answer: A RS232 to USB converter is required to convert the serial data into computer readable form.

288. After initializing UART how the data flow will take place?

Answer: Once the module is initialized whatever value is loaded into the register TXREG will be transmitted through UART, but transmission might overlap. Hence, we should always check for the Transmission Interrupt flag TXIF.

289. How to receive data using UART?

Answer: When a data is received by the UART module it picks it up and stores it up in the RCREG register. We can simply transfer the value to any variable and use it.

290. Give the detail of ADC used in PIC16F877?

Answer: 10-bit, up to 8-channel Analog-to-Digital Converter (A/D)

291. Why do we use ADC in PIC16F877?

Answer: The sensors like temperature sensor, flux sensor, pressure sensor, current sensors, voltage sensors, gyroscopes, accelerometers, distance sensor, produces an analog voltage of 0V to 5V based on the sensors reading, to read it through controller ADC is required.

292. What types of ADC are available with PIC16F877?

Answer: There are many types of ADC available and each one has its own speed and resolution. The most common types of ADCs are flash, successive approximation, and sigma-delta. The type of ADC used in PIC16F877A is called as the **Successive approximation ADC**.

293. What do you mean by Successive Approximation ADC?

Answer: The SAR ADC works with the help of a comparator and some logic conversations. This type of ADC uses a reference voltage (which is variable) and compares the input voltage with the reference voltage using a comparator and difference, which will be a digital output, is saved from the **Most significant bit (MSB)**.

294. How to use the ADC on our PIC16F877 MCU.

Answer: The PIC has 10-bit 8-channel ADC. This means the output value of our ADC will be 0-1024 (2^10) and there are 8 pins (channels) on our MCU which can read analog voltage. The value 1024 is obtained by 2^10 since our ADC is 10 bits.

295. How many registers are there to deal the ADC of PIC16F877.

Answer: The A/D module has four registers which has to be configured to read data from the Input pins.

a. A/D Result High Register (ADRESH)
b. A/D Result Low Register (ADRESL)
c. A/D Control Register 0 (ADCON0)
d. A/D Control Register 1 (ADCON1)

296. How to initialize the ADC with PIC16F877.

Answer: Inside the void main() we have to initialize our ADC by using the ADCON1 register and ADCON0 register.

297. What steps are to be taken to read a value from ADC.

Answer: To read an ADC value the following steps has to be followed.

i. Initialize the ADC Module
ii. Select the analog channel
iii. Start ADC by making Go/Done bit high
iv. Wait for the Go/DONE bit to get low
v. Get the ADC result from ADRESH and ADRESL register

298. Most of the instructions in PIC family are of how many cycle?

Answer: Single cycle

299. Who developed the architecture of AVR

Answer: Alf-EgilBogen and VegardWollan, from Norvegian Institute of Technology (NTH) developed the basic architecture of AVR.

300. What are the standard features of AVR?

Answer: Some of the features of AVR are:
- On-chip program ROM
- Data RAM
- Data EEPROM
- Timers

- Input-Output ports
- ADC
- PWM
- I2C
- SPI
- USART
- USBN
- CAN

301. What is the maximum size of RAM in AVR?

Answer: AVR support maximum of 64K bites of data RAM.

302. How many input/output pins does AVR has?

Answer: AVR comes with 3 to 86 input-output pins.

303. What are the different peripherals supported by AVR?

Answer: The different peripherals supported by AVR are:
- ADC
- PWM
- I2C
- SPI
- USART
- USBN
- CAN

304. Name the different AVR families

Answer: Different types of AVR families are:
- Classic AVR (AT90Sxxxx)
- Mega AVR (ATmegaxxxx)
- Tiny AVR (ATtinyxxxx)

305. Discuss the characteristics of Mega AVR family?

Answer: The characteristics of Mega AVR family are:
- 4K to 256K bites of program memory
- 28 to 100 pins package

- Extensive peripheral set
- Rich instruction set

306. Discuss the characteristics of Tiny AVR family

Answer: The characteristics of ting AVR family are:

- 1K to 8K bites of program memory
- 8 to 28 pins package
- Limited peripheral set
- Limited instruction set

307. Discuss the General-Purpose Resistor (GPR) in AVR

Answer: AVR consist of 32 GPR located at the lowest of memory address.

308. Does AVR have accumulator?

Answer: AVR does not have dedicated accumulator but all 32 General Purpose Resistors can act as accumulator and can be used for arithmetic and logical operations.

309. Compare SRAM with EEPROM in AVR chips

Answer: EEPROM does not lose the stored data when the power is off while SRAM loses the data. So EEPROM is used to store the data that is rarely changed and need to be preserved even after power off.

310. Define AVR Carry flag

Answer: When there is a carry out from D7 the carry flag is set.

311. Define AVR Zero flag

Answer: When the result of arithmetic and logic operation is zero, zero flag is set for known zero result zero flag is reset.

312. Define AVR Negative flag

Answer: While representing signed binary numbers D7 is used as signed bit. If negative flag is 1 result of arithmetic operation is negative.

313. Define AVR Overflow flag

Answer: Whenever the result of the arithmetic operation of signed numbers is sufficiently large leading higher bit of the result to overflow into signed bit the overflow flag is set.

314. Define AVR Sign bit?

Answer: The sign bit is affected as a result of Exclusive-ORing of N and V flags.

315. Explain AVR Half Carry flag?

Answer: Half carry flag is set when there is a carry from D3 to D4 in addition or subtraction operation.

316. Give the Conditioner flags in AVR?

Answer: Carry (C), Zero (Z), Negative (N), Overflow (V), Sign (S), Half carry (H) flags.

317. How the flags are placed/arranged in AVR status resistors

Answer: Status register is:

I	T	H	S	V	N	Z	C

318. Show the status of C, H and Z flags after the addition ref 0 38 &0 2F

Answer: C=0, H = 1, Z = 0

319. Show status of C, H and Z after addition of 0 88 &0 93

Answer: C=1, H = 1, Z = 0

320. How Hex number is represented in AVR?

Answer: Hex number is AVR are represented in two ways:

a. By putting 0x in front of number like 0x38
b. By putting $ in front of number like $38

321. How Binary numbers are represented in AVR?

Answer: Binary number is represented in AVR by putting 0b or 0B in front of number like 0b10011011 or 0B10011011

322. How Decimal numbers are represented in AVR?

Answer: Decimal numbers are used directly in AVR without putting anything before or after the number.

323. How ASCII numbers are represented in AVR?

Answer: To represent the ASCII number, the number is put between single quotes like 'A'.

324. Explain .EQU assembler directive.

Answer: EQU is used to give definition to a constant or a fixed address like .EQU COUNT 0x45

325. Explain .SET assembler directive

Answer: .SET is used to give definition to a constant or a fixed address similar to .EQU.

326. Explain .ORG assembler directive

Answer: .ORG directive is used to tell the compiler the beginning address for the code written.

327. Give the rules for labels in Assembly Language for AVR.

Answer: Rules for label in assembly language are:

- Each label must be unique
- Label consists of alphabetic letters both uppercase and lowercase
- Label can also have digits from 0 to 9
- Label can also have digits special characters like @,#,&
- First letter of the label needs to be alphabet.
- Reserved word in assembly should not to be used as label

328. Give the structure of Assembly Language for AVR.

Answer: Assembly language consists of following fields:

[label:]   mnemonic   [operands]   [;comments]

- **Label:** It is used to refer a line of code with the name
- **Mnemonic:** These are the actual instruction in assembly code
- **Operands:** Operands are the data or memory locations on which the action is to be performed. Mnemonic along with operands perform the real work
- **Comment:** It is the short description for the line of code

329. Give the details of .lst file

Answer: The .lst file gives the instruction used in the program and the program memory occupied by the code.

330. What is the effect/advantage of having wider program counter inn AVR?

Answer: With the wider program counter, more memory locations can be accessed by the CPU.

331. Give the size of maximum memory that can be accessed with 16-bit PC

Answer: With 16 – bit of program counter, maximum of 64K bytes of memory can be accessed.

332. What is the size of each ROM location in AVR?

Answer: The size of each location of AVR is 2 bytes or 16-bits. For example, 2 KB of memory will have 2048 bytes and in AVR it will be having 1024 (2048/2) locations.

333. Find ROM memory address for ATmega 16 with 16 KB ROM

Answer: With 16 KB of memory, the number of locations will be 16384 (16x1024). But in AVR as each location is of 2 bytes, it will be having 8092 (16384/2) locations which gives address range from 0000 to $1FFF.

334. What is the size of internal bus in AVR between code ROM and CPU?

Answer: The internal bus that runs between code ROM and CPU is 16-bit byte.

335. What is Harvard Architecture?

Answer: In Harvard Architecture there are two types of memories i.e. Code memory and data memory. For code and data memories separate buses are also implemented.

336. What is Von-Newman Architecture?

Answer: In Von-Newman Architecture, there is only single memory which is divided into code segment and data segment and there is only one set of data and address.

337. Give the features of RISC architecture.

Answer:

The various features of RISC architecture are:

- Fixed instruction size
- Large numbers of resistors
- Small instruction set with only basic instructions
- 95% of the instructions executes with only one clock cycle
- With small instruction set, instructions are implemented using hard wire method taking only about 10% of transistors against 40-60% of transistors employed on CISC architecture
- RISC uses load/store architecture

338. What is load/store architecture?

Answer: RISC uses load/store architecture in which the data from memory is to be loaded to CPU resistors for operation and result is stored back to the memory. While in CISC data can be manipulated while it resides in memory.

339. Why C is preferred over assembly for programming RISC architecture microcontroller?

Answer: The assembly language instruction set is different for different microcontroller because of which the code written in assembly language written for one microcontroller cannot be used for another microcontroller. The assembly programming needs extensive knowledge of underlying hardware. While C programming for microcontroller is portable with little change, easy to read and need very little knowledge underlying hardware.

340. Give the different 'C' data types in embedded C and their Range.

Data Type	Size in bits	Data Range
Char	8-bit	-128 to +127
unsigned char	8-bit	0 to 255
Int	16-bit	-32,768 to +32,767
unsigned int	16-bit	0 to 65,535
Long	32-bit	-2,147,483,648 to +2,147,483,647

unsigned long	32-bit	0 to 4,294,967,295
Float	32-bit	+/- 1.175e-38 to +/-3.402e38
double	32-bit	+/- 1.175e-38 to +/-3.402e38

341. Why there are more than one Ground pins in AVR?

Answer: In order to reduce the noise i.e. ground bounce at high frequency the chips with 40 pins or more generally have multiple grounds.

342. What is the function of AREF pin in ATmega32?

Answer: AREF pin is the analog reference pin for internal ADC of AVR. It is used to define range of input analog voltage for AVR ADC.

343. What is the range of crystal oscillator that can be connected to ATmega32?

Answer: AVR ATmega32 can be connected with crystal from 0 Hz to 16 MHz.

344. What is the voltage range for ATmega32L?

Answer: ATmega32L is designed to operate from 2.7 V to 5.5 V.

345. How many I/O port does ATmega32 has?

Answer: ATmega32 has 4 ports each of 8 bits. The ports of ATmega32 are named as PORT A, PORT B, PORT C, PORT D.

346. On reset I/O ports of ATmega32 is initialized as input or output?

Answer: The content of register DDRx is 0x00 which indicates that I/O ports of ATmega32 are initialized as input port upon reset.

347. Why unsigned character V-variable types is preferred for 8-bit microcontrollers?

Answer: Unsigned char C variable is preferred because it is 8-bit wide which is same as ALU data length and size of single memory location in 8-bit microcontrollers.

348. Which bits of port C are also having I2C bus?

Answer: Port C bit 0 (PC0) and port C bit 1 (PC1) are the port C bits which are also having I2C bus functionality. PC 0 is SCL (serial clock) and PC 1 is SDA (serial data).

349. Which bits of port B are also having SPI bus?

Answer: Port B bit 4 (PB4), port B bit 5 (PB5), port B bit 6 (PB6) and port B bit 7 (PB7) are the port B bits which are also having SPI bus functionality. PB 4 is SS (slave select), PB 5 is MOSI (master out slaving) PB 6 is MISO (master in slave out) and PB 7 is SCK (serial clock).

350. Discuss MOSI.

Answer: MOSI stands for Master Out Slaving In. This is output pin of microcontroller in SPI communication.

351. Discuss MISO.

Answer: MISO stands for Master In Slave Out. This is input pin of microcontroller in SPI communication.

352. Which pins of ATmega32 is having UART serial bus?

Answer: Pin number 14 and pin number 15 are dedicated to USART serial bus. Pin number 14 is RxD and pin number 15 is TxD.

353. Name the various serial bus protocols supported by ATmega32.

Answer: The various serial protocols supported by ATmega32 are USART SPI I2C.

354. What are the different clock sources in ATmega32?

Answer:

The various clock sources for ATmega32 are:
- External RC oscillator
- External clock
- External oscillator
- Low-frequency crystal oscillator
- Calibrated RC oscillator

355. How are the CKSEL3 to CKSEL0, select the clock frequency using internal RC Network?

Answer:

CKCEL3210	Frequency (MHz)
0001	1
0010	2
0011	4
0100	8

356. What is the purpose of fuse bits in ATmega32?

Answer: The fuse bits in AVR are used to select certain features of AVR that can reduce cost by eliminating need of external components.

357. How are the CKSEL3 to CKSEL0, select the clock frequency using external RC Network?

Answer:

CKCEL3210	Frequency (MHz)
0101	<0.9
0110	0.9-3.0
0111	3.0-8.0
1000	8.0-12.0

358. How the machine cycle for AVR is calculated?

Answer: Machine cycle of AVR is calculated by directly converting the frequency of oscillator selected into time.

359. How the machine cycle of AVR is different from 8051?

Answer: In 8051, machine cycle is calculated first by dividing crystal oscillator by 12 and then converting the obtained frequency into time while machine cycle of AVR is calculated by directly converting the frequency of oscillator selected into time.

360. Compare AVR, PIC & 8051.

Answer:

- Machine cycle of AVR is calculated by directly converting the frequency of oscillator selected into time

- In PIC, machine cycle is calculated first by dividing crystal oscillator by 4 and then converting the obtained frequency into time
- In 8051, machine cycle is calculated first by dividing crystal oscillator by 12 and then converting the obtained frequency into time

361. Find the instruction cycle timing for ATmega32 for 4 MHz crystal.

Answer: Instruction cycle timing is 1/4 (MHz) = 250ns

362. Find the instruction cycle timing for ATmega32 for 16 MHz crystal.

Answer: Instruction cycle timing is 1/16 (MHz) = 62.5ns

363. Why start up delay time is required in AVR?

Answer: In AVR, startup time is required to allow CPU clock source and voltage to get stabilized.

364. What is start up time in AVR?

Answer: Startup time is the delay which is provided to the AVR after applying the power during which AVR's clock and voltage source are stabilized.

365. How the start-up delay is selected in AVR?

Answer: In AVR, startup delay is selecting using SUT1, SUT0 and CKSEL0 fuse bits.

366. What is the provision given in ATmega for power source fluctuation?

Answer: For taking care of power source fluctuation in ATmega brown-out deduction (BOD) is provided.

367. How Brown-out detector (BOD) takes care of power source fluctuation?

Answer: BOD take cares of power source fluctuation by comparing Vcc with the BOD level and if Vcc level is detected BOD level it resets the chip.

368. Give the different Hex file format produced by AVR studio.

Answer: The Hex file formats which are produced by AVR studio are

File Extension	Format Name	Maximum ROM Address
.hex	Extended Intel Hex file	20-bit address
.mot	Motorola S-record	32-bit address
.gen	Generic	24-bit address

369. How many Registers are associated with each port of AVR?

Answer: Registers which are associated with each port of AVR are:

- DDRx
- PORTx
- PINx

Where x is the specific port

370. What is DDR stands for?

Answer: DDR stands for data direction resistor

371. What is the role of DDR register in AVR?

Answer: DDR resistor in AVR is used to define the port as input or output. Writing 1 to the DDR port will define the port pin as output and writing 0 to the DDR port will define the port pin as input.

372. What is the PIN register role in inputting date?

Answer: In order to read the data present at the port pins of AVR pin register is required to be read.

373. What are the different states of AVR pins according to values of PORT & DDR?

Answer: The different states of AVR pins are:

PORTxDDRx	0	1
0	Input and high impedance	Out0
1	Input and pull-up	Out1

374. Which pins of ATmega32 are connected with PORT A, PORT B, PORT C & PORT D?

Answer:
- PORT A is from pin number 33 to 40
- PORT B is from pin number 1 to 8
- PORT C is from pin number 22 to 29
- PORT D is from pin number 14 to 21

375. How to define PORT D as input?

Answer: By writing DDRD with value 0 (decimal) or 0x00 (hexadecimal) PORTD is defined as input.

376. How to define PORT D as output?

Answer: By writing DDRD with value 255 (decimal) or 0xFF (hexadecimal) PORTD is defined as output.

377. Write C-program to send 00-55 to PORT B.

Answer:
```
#include <avr/io,h>
Int main (void)
{
Unsigned char z;
DDRB = 0xFF;
for (z = 0; z <= 255; z++)
PORTB = z;
return 0;
}
```

378. Which heads file is used in AVR C-coding?

Answer: <avr/io,h> is the header file used in AVR C-coding.

379. What is the difference between char & unsigned char in C data types?

Answer: The range of char is from -128 to +127 while the range of unsigned char is from 0 to 255.

Exercise 125

380. What is the difference between int & unsigned int in C data types?

Answer: The range of int is from -32,768 to +32,767 while the range of unsigned int is from 0 to 65,535.

381. What are the different ways of creating time delays in AVR C programming?

Answer: Different ways of creating time delays in AVR C programming are using simple for loop, using predefined C function and using AVR timers.

382. What is the hardware way of manipulating time delays in AVR?

Answer: The delays in AVR can be changed by changing the crystal (XTAL) hardware.

383. Which header file is used in AVR for creating time delay?

Answer: <util/delay.h> is the header file used in AVR for creating delay.

384. How to generate delay of 50ms using <util/delay.h> for header?

Answer: _delay_ms(50) is the function with argument is to be called for creating a delay of 50 ms using <util/delay.h>

385. How to test equality of 2-Number using logical operators in AVR?

Answer: XOR(^) is the logical operator to be used for testing the equality of two numbers.

386. How to set a particular bit of 8-bit data using logical operators?

Answer: For setting a particular bit of 8-bit data that particular bit needs to be ORed with 1 and rest of the bits need to be ORed with 0.

387. How to reset a particular bit of 8-bit data using logical operators?

Answer: For re-setting a particular bit of 8-bit data that particular bit needs to be ANDed with 0 and rest of the bits need to be ANDed with 1.

388. What is the meaning of 30H>>4?

Answer: The meaning of 30H>>4 is to shift the data 30H by 4 position towards left in 8-bit resistor.

389. How to read the status of PB5 in C for ATmega32?

Answer: PINB & (1<<5) is the instruction to read the status of PB5 in C for ATmega32.

390. How many timer does ATmega32 has?

Answer: AVR ATmega32 has 3 timers:Timer0, Timer1, Timer2

391. Give the bit wise length of ATmega32 timers.

Answer: In AVR ATmega32 timers0 and timer2 are of 8-bit while timer1 is of 16-bit.

392. Define timer.

Answer: Timer needs clock pulse to tick. The source of clock pulse can be internal or external. If the clock pulse is internal, timer is used to generate the delays, so it is called as timer.

393. Define Counter.

Answer: Timer needs clock pulse to tick. The source of clock pulse can be internal or external. If the clock pulse is external, timer is used to count the external pulses and so it is called as counter.

394. Name the Registers and bits associated with each timer of ATmega32.

Answer:

Following registers and bits are associated with each timer of ATmega32.

- TCNTn-timer/counter resistor (TCNT0, TCNT1, TCNT2)
- TOVn-timer overflow flag (TOV0, TOV1,TOV2)
- TCCRn-timer/counter control resistor (TCCR0, TCCR1, TCCR2)
- OCRn-output compare resistor (OCR0, OCR1, OCR2)
- OCFn-output compare flag (OCF0, OCF1, OCF2)

395. Explain TCNTn Register.

Answer: TCNT resistor is a counter resistor which counts up for each pulse and on reset it is loaded with 0.

396. Explain TUVn flag of timer in ATmega32.

Answer: The TUVn flag indicate the completion of work by timer, which is done when the timer overflows, when the timer completes its work i.e. when it overflows TOV flag is raised to 1.

397. Explain TCCRn register for timer in ATmega32.

Answer: TCCRn resistor is used to setup the mode of the timer to function as timer or counter.

398. Explain OCRn register for timer in ATmega32.

Answer: The OCRn resistor is used to hold a value which can be compared with TCNTn and when the content of OCRn and TCNTn are equal OCFn flag is said to 1.

399. What is the length of TCNT register for T0, T1 & T2 in ATmega32?

Answer: The resistor TCNT0 for timer0 is of 8-bit, TCNT1 for timer1 is of 16 bit, TCNT2 for timer2 is of 8 bit.

400. Find TCCR0 value in normal mode without any pre-scalar using crystal oscillator as clock source.

Answer: Hex data for TCCR0 will be 0X00 with normal mode without any pre-scalar using crystal oscillator as clock source.

401. Calculate timer clock frequency and period for 10MHz & 16MHz crystal, without any pre-scalar.

Answer:

- F = 10MHz and T = 1/10MHz = 0.1us
- F = 16MHz and T = 1/10MHz = 0.0625us

402. What are the steps involved in programming timer 0 in normal mode of ATmega32?

Answer:

Steps for programming timer 0 in normal mode:

- TCNT0 resistor is loaded with initial count value
- Load TCCR0 resistor to select mode (8-bit or 16-bit) and pre-scalar option
- As soon as the clock source is selected the counter will start counting for each tick and the timer will be increment by 1

- By monitoring timer overflow flag (TOV0) timer overflow status is checked and if flag is raised come out of the loop
- Stop the timer by disconnecting the clock
- Clear the TOV0 flag go back to step 1

403. How to find a value to be loaded to timer resistor TCNT0 of ATmega32?

Answer: In order to find the value to be loaded to consider an example of delay using 135 clocks of 0.125us so

- Enter the value 135 into the scientific calculator
- Select the hexadecimal mode to convert 135 into hexadecimal, which is 0x87
- Select +/- to give -135 decimal, which is 0x79
- The lower two digits i.e. 79 is the value to be loaded to TCNT0 resistor

404. What value to be loaded to TCNT0 to generate a square wave of 16 KHz XTAL = 8MHz?

Answer: The steps involved are:

1. Convert the frequency of square wave into time, T = 1/16 KHz = 62.5us, which is time period of square wave
2. Divide the time period obtained by 2 for the high and low portion of square wave which comes out as 31.25us
3. Divide the number obtained in step 2 with the machine cycle of AVR @XTAL = 8 MegaHz i.e. 0.125us (31.25/0.125 = 250)
4. Subtract the number obtained in step 3 from 256 to obtain the value to be loaded in TCNT0, 256-250 = 6 or 0x06

405. Give the value for TCNT0 to get largest possible delay with crystal of 8 MegaHz.

Answer: The largest possible delay with TCNT0 will be 32us (256x0.125) by loading TCNT0 with 0x00.

406. What are the possible pre-scalar values for ATmega32 timer 0?

Answer: No pre-scalar, 1/8, 1/64, 1/256, 1/1024 are the possible pre-scalar ATmega32 timer 0.

407. Explain CTC mode of timer 0 programming in AVR.

Answer: CTC mode stands for clear timer 0 on compare match. In this OCR0 resistor is used and loaded with the value. TCNT0 resistor of timer 0 will increment by 1 for each clock tick and when its value become equal to OCR0, timer is cleared and OCF0 flag is raised.

408. List out the various capabilities of Timer1.

Answer:

The various capabilities of timer1 are:

- Timer1 is 16bit timer split into TCNT1L (timer1 low byte) and TCNT1H (timer1 high byte) resistors
- Timer1 consist of 2 control resistors namely TCCR1A (timer/counter 1 control resistor) and TCCR1B
- TOV1 is timer overflow flag bit and is said to 1 when timer overflow occurs
- Timer1 pre-scalar options are 1:1, 1:8,1:64, 1:256 and 1:1024
- Timer1 has 2 OCR resistors, OCR1A and OCR1B, with each OCR resistors having its own flags.

409. What is Normal mode of operation for Timer1?

Answer: In normal mode, timer counts up till it reaches 0xFFFF and then roll over to 0x0000 and sets the flag TOV1.

410. What is CTC mode of operation for Timer1?

Answer: In CTC mode, the resistor TCNT1 content are incremented till they become equal to content of OCR1A resistor. When the content of two resistors are equal the timer resistor is cleared by the next clock pulse.

411. How to generate large time delay using timer?

Answer: The large time delay can be generated by using the timer pre-scalar options available from 8 to 1024.

412. What are the factors on which time delays are dependent in AVR?

Answer:

The various factors on which the time delay is dependent in AVR are:

- Crystal frequency
- Timer's resistor (8-bit or 16-bit)
- Timer pre-scalar options (available from 8 to 1024)

413. How interrupts are executed in AVR?

Answer:

Following steps are involved in the execution of the interrupt:

- When interrupt occurs, microcontroller finishes the execution of current instruction.
- Thereafter microcontroller jumps to a fixed location in the memory obtained from Interrupt Vector Table (IVT).
- Microcontroller starts the execution of the instructions present at the address obtained from IVT and is called Interrupt Service Routine (ISR)
- Upton encountering the RETI instruction microcontroller return to the address in the memory where it was interrupted.

414. Define Interrupt.

Answer: In interrupt whenever any device needs services of microcontroller it sends signal to signified it.

415. Define polling?

Answer: In polling, microcontroller continuously monitors the device for its status and the requirement of service by the device from the microcontroller.

416. What is IVT interrupt vector table?

Answer: Interrupt vector table is the table which contains the addresses of the various interrupt service routine for the supported interrupt.

417. Give the ROM locations for External interrupt.

Answer:

The ROM locations for external interrupt are:

- 0x0002 for external interrupt request 0
- 0x0004 for external interrupt request 1
- 0x0006 for external interrupt request 2

418. Give the ROM location for Timer/counter interrupts.

Answer:

The ROM location for Timer/counter interrupts are:
- 0x0008 for timer/counter2 compare match
- 0x000A for timer/counter2 overflow
- 0x000C for timer/counter1 capture event
- 0x000E for timer/counter1 compare match A
- 0x0010 for timer/counter1 compare match B
- 0x0012 for timer/counter1 overflow
- 0x0014 for timer/counter0 compare match
- 0x0016 for timer/counter0 overflow

419. Give the ROM location for Serial Bus interrupts in AVR.

Answer:

The ROM location for Serial Bus interrupts are:
- 0x0018 for SPI transfer complete
- 0x001A for USART, Receive complete
- 0x001C for USART, data register empty
- 0x001E for USART transmit complete

420. What are the different sources of interrupts in AVR ATmega32?

Answer:

Different sources for interrupt in AVR are:
- Two interrupts for timer
- Three external hardware interrupts
- Three interrupts for USART serial communication
- One SPI interrupt
- One ADC interrupt

421. How interrupts are enabled and disabled in AVR?

Answer: In AVR, interrupts are enable setting the bit I of register SREG (Status Register) and disabled by resetting the bit I of SREG (Status Register).

422. What are the steps involved in enabling the AVR interrupt?

Answer: First D7 (I) of SREG register is made high to enable the interrupts. There after individual interrupt in enable by setting the Interrupt Enable (IE) flag bit of that interrupt.

423. What is the difference between RET and RETI instruction in AVR?

Answer: RET instruction is used to return from an assembly subroutine but RETI instruction is used to return from an Interrupt subroutine. If RET is used in place of RETI, it will block the interrupt as RETI also perform the task of setting up the 'I' flag indicating that interrupt is over.

424. What is "Compare match timer flag" in AVR?

Answer: In compare match timer mode, OCF flag i.e. flag for Compare match timer is se t when the OCR register contents matches with the TCNT register.

425. On which I/O Port pin external interrupt are located?

Answer: The external interrupt INT0, INT1 and INT2 are located on PORTD2, PORTD3 and PORTB2 respectively.

426. By default, what is the signal required to evoke INT0 external interrupt?

Answer: By default, INT0 interrupt is low level triggered.

427. What is difference between level triggered and edge triggered interrupts?

Answer: Interrupts which are activated by HIGH or LOW signal are level triggered interrupt while interrupts which are triggered by the HIGH to LOW or LOW to HIGH transition signal are edge triggered interrupts.

428. Which of the external interrupts are level triggered and edge triggered in AVR ATmega32?

Answer: In AVR ATmega32 only INT2 is level triggered while INT0 and INT1 can be level or edge triggered.

429. How to select the trigger option for external interrupt INT0 & INT1?

Answer: INT0 and INT1 are low-level triggered on reset and MCUCR register bits define the trigger options for INT0 and INT1.

430. What is interrupt priority in AVR?

Answer: If two interrupts are activated simultaneously, then this type of the clash is resolved by the interrupt priority. In this situation, interrupt with the higher priority will be serviced first and then the lower priority interrupt will be serviced.

431. How the interrupt priority is related to memory address in Interrupt Vector Table (IVT) in AVR.

Answer: Interrupts having the lower address in Interrupt Vector Table is having higher priority. For example, eternal interrupt 0 address in Interrupt Vector Table is 2 and that of external interrupt 2 in 6 thus external interrupt 0 has higher priority over external interrupt.

432. How many resistors are associated with AVR USART?

Answer:

There are five resistors associated with AVR USART are:

- UDR-USART data resistor
- UCSRA, UCSRB, UCSRC- USART control status resistor
- UBRR- USART baud rate resistor

433. What is the relation between FOSC and the value to be loaded to UBRR in AVR?

Answer: Desired baud rate = Fosc/(16(X+1)) where X is the value to be loaded to UBRR

434. What value should be loaded to UBBR with FOSC as MHz for 9600 Band Rate in AVR?

Answer: For 8 MHz crystal

X = (8MHz/16(desired baud rate))-1

Using the above formula for 9600 baud rate value of X is 51.08 i.e. 51 or OX33

435. How the error in Baud Rate calculation is obtained for AVR?

Answer: The formula for calculating error in the baud rate is:

Error = (calculated value of the UBRR – integer part)/integer part

436. Find the % error in Baud Rate for 9600 Band Rate at 8 MHz crystal.

Answer: Using the formula in for error in baud rate the percentage error in setting baud rate at 9600 for 8 MHz crystal will be 0.16 i.e. (51.08 − 51)/51

437. Define step size in ADC.

Answer: The step size is defined as smallest chain in the input analog voltage that can be detected by ADC.

438. What is the role of V_{ref} in AVR ADC?

Answer: The voltage connected to V_{ref} dictates the step size of ADC e.g. for 8-bit ADC step size is $V_{ref}/256$.

439. Calculate the step size with V_{ref} connected to 2.56 V for 8-bit ADC.

Answer: For 8-bit ADC, step size is $V_{ref}/256$ i.e. 2.56/256 which gives step size as 10 mV.

440. What is the bitwise size of ADC in ATmega32?

Answer: AVR ATmega32 consists of 10-bit ADC, digital count ranging from 0 to 1024.

441. How many channels of ADC are supported by ATmega32?

Answer: AVR ATmega32 consists of 8 channel 10-bit ADC.

442. Which register holds the converted data of ATmega32 ADC?

Answer: Register ADCL (ADC result low) and ADCH (ADC result high) holds the ADC converted data.

443. How the source of V_{ref} is selected in AVR ATmega16?

Answer: In AVR V_{ref} source is selected by programming REFS1 and REFS0 bits of ADMUX register as per following table:

REFS1	REFS0	V_{ref}	
0	0	AREF pin	Set externally
0	1	AVCC pin	Same as VCC
1	0	Reserved	----
1	1	Internal 2.56 V	Fixed regardless of VCC

444. How the ADC channel is selected in AVR ATmega32?

Answer: In AVR ATmega32, ADC channel is selected by programming MUX0, MUX1, MUX2, MUX3 and MUX4 bits of ADMUX register as per following table:

MUX4...0	ADC channel
00000	ADC0
00001	ADC1
00010	ADC2
00011	ADC3
00100	ADC4
00101	ADC5
00110	ADC6
00111	ADC7

445. How the ADC pre-scalar is selected in AVR ATmega32?

Answer: ADC pre-scalar is selected by programming ADPS0, ADPS1 and ADPS2 bits of ADCSRA (ADC status and control) register as per following table:

ADPS2	ADPS1	ADPS0	ADC clock (CK)
0	0	0	Reserved
0	0	1	CK/2
0	1	0	CK/4
0	1	1	CK/8
1	0	0	CK/16
1	0	1	CK/32
1	1	0	CK/64
1	1	1	CK/128

446. In ATmega32, 32 signifies what?

Answer: ATmega32 the number 32 signifies 32K bytes of flash memory.

447. Which port can be used for interfacing for the LED in AVR?

Answer: Any port out of RA, RB, RC or RD can be used for interfacing of LED.

448. LED's can be used either sink or source mode?

Answer: Any mode can be used but sink mode is preferred.

449. Does it necessary to use serial resistor with LED and if yes what value?

Answer: Generally, 100-250 Ohms resistance can be used. But if we use higher value resistance than LED brightness will reduce.

450. How to control the direction of rotation for DC motor?

Answer: By changing the power voltage connected to the motor terminals direction of rotation of DC motor is controlled.

451. How to control the speed of DC motor?

Answer: By varying the voltage connected to DC motor speed of DC motor is controlled.

452. What is the digital way of controlling the speed of DC motor?

Answer: By applying the PWM signal to DC motor through the motor driver speed of the motor is controlled.

453. What is the function of H – bridge?

Answer: H – bridge is used in interfacing of DC motor for its direction controlled. It's a 4 – switch circuit and by controlling the switched ON/OFF position direction of rotation for DC motor is controller.

454. Name any motor driver used for interfacing of DC motor with microcontroller

Answer: L293D motor driver is used to interface DC motor with microcontroller.

455. Name any motor driver used for interfacing of stepper motor with microcontroller.

Answer: ULN2003A motor driver is used to interface stepper motor with microcontroller

456. What is the difference between half step and full step sequence?

Answer: Full step sequence gives the more precise motor movement control as compared to half step sequence.

457. What is the general application of servo motor?

Answer: Servo motors are used where precise angular movement need to be controlled and high static torque is required.

458. How the angle of movement in servo motor is controlled?

Answer: In servo motor, PWM signal is used to control the angle of movement in servo motor is controlled.

459. List three sources of possible errors in instruments.

Answer:

Three sources of possible errors in instruments are:

a. Gross Error
b. Systematic
c. Random errors.

460. Define Instrumental error.

Answer: These are the errors inherent in measuring instrument because of their mechanical structure.

461. Define limiting error.

Answer: Components are guaranteed to be within a certain percentage of rated value. Thus, the manufacturer has to specify the deviations from the nominal value of a particular quantity.

462. Define probable error.

Answer: Probable error is defined as $r = \pm 0.6745s$ where s is standard deviation. Probable error has been used in experimental work to some extent in past, but standard deviation is more convenient in statistical work.

463. Define Environmental error

Answer: Environmental error are due to conditions in the measuring device, including conditions in the area surrounding the instrument, such as the effects of changes in temperature, humidity.

464. Define arithmetic mean.

Answer: The best approximation method will be made when the number of readings would give the best result.

465. Define average deviation.

Answer: By definition, average deviation is the sum of absolute values of the value deviations divided by the number of reading.

466. Define transducer and give an example.

Answer: Transducer is a device which converts one form of energy into other form of energy.

A thermocouple converts heat energy into electrical voltage.

467. Classify transducer.

Answer:

Transducer are classified as:
- Primary and secondary transducers
- Active and passive transducers
- Analog and digital transducers
- Transducers and inverse transducers

468. What is primary transducer?

Answer: Bourdon tube acts as a primary transducer senses the pressure and converts the pressure into displacement. No output is given to the input of the bourdon tube. So, it is called primary transducer. Mechanical device can act as a primary transducer.

469. What is secondary transducer?

Answer: The output of the Bourdon tube is given to the input of the LVDT. There are two stages of transduction, firstly the pressure is converted into a displacement by the Bourdon tube then the displacement is converted into analog voltage by LVDT. Here LVDT is called secondary transducer. Electrical device can act as a secondary transducer.

470. What is passive transducer?

Answer: Transducer which cannot work in the absence of external power, and it is called a passive transducer.

Example: capacitive, inductive, resistance transducers.

471. What is active transducer?

Answer: Transducer which can work in the absence of external power, and it is called an active transducer.

Example: velocity, temperature, light can be transduced with the help of an active transducer.

472. What is analog transducer?

Answer: Analog transducers convert the input quantity into an analog output which is a continuous function of time. Thus, a strain gauge, an LVDT, a thermocouple or a thermistor may be called analog transducer, as they give an output which is a continuous function of time.

473. Give the classification of units.

Answer:

The classification of units:

- Absolute units
- Fundamental and derived units
- Electromagnetic units
- Electrostatic units

474. Define static characteristics.

Answer: Static characteristics of a measurement system are, in general, those that must be considered when the system or instrument is used to measure a condition not varying with time.

475. Mention different types of static characteristics.

Answer:

Different types of static characteristics are:

- Accuracy
- Sensitivity
- Reproducibility
- Drift
- Static error and
- Dead zone

476. What are dynamic characteristics?

Answer: Many measurements are concerned with rapidly varying quantities and, therefore, for such cases we must examine the dynamic relations which exist between the output and the input. This is normally done with the help of differential equations. Performance criteria based upon dynamic relations constitute the Dynamic Characteristics.

477. Mention different type's dynamic characteristics?

Answer:

Different types of dynamic characteristics are:
- Zero- order transducers
- First – order transducers
- Second-order transducers
- Higher-order transducers

478. Compare accuracy and precision.

Answer: Accuracy is the closeness to true value whereas precision is the closeness amongst the readings.

Precision is the degree of closeness with which a given value may be repeatedly measured.

479. Define resolution.

Answer: When the input to a transducer is increased slowly from some non-zero arbitrary value, the change in output is not detected at all until a certain input increment is exceeded. This increment is defined as the resolution.

480. Define hysteresis.

Answer: When the input to a transducer which is initially at rest is increased from zero to full-scale and then decreased back to zero, there may be two output values for the same input. Hysteresis effects can be minimized by taking readings corresponding to ascending and descending values of the input and then taking their arithmetic average.

481. What is range and span?

Answer: The range of the transducer is specified as from the lower value of input to higher value of input.

The span of the transducer is specified as the difference between the higher and lower limits of recommended input values.

482. What is potentiometer?

Answer: A variable resistor is a potentiometer, or simply a POT, (a resistive potentiometer used for the purposes of voltage division is called a POT) consists of a resistive element provided with a sliding contact. The POT is a passive transducer.

483. What is piezo-electric effect?

Answer: A piezo-electric material is one in which an electric potential appears across certain surfaces of a crystal if the dimensions of the crystal are changed by the application of the mechanical force.

484. What is digital transducer?

Answer: Digital transducer converts input quantity into an electrical output which is in the form of pulses.

485. What is piezo-electric transducer?

Answer: Piezo-electric transducer convert pressure or force into electrical charge. These transducers are based upon the natural phenomenon of certain non-metal and di-electric components.

486. What are the suitable materials for piezo electric transducer?

Answer: Primary quartz, Rochelle salt, ammonium di-hydrogen phosphate (ADP), and ceramics with barium titanate, di-potassium tartrate, potassium di-hydrogen phosphate and lithium sulfate are the suitable material for piezo electric transducer.

487. What is 'd' coefficient?

Answer: It gives the charge output per unit force input (or charge density per unit pressure) under short circuit condition, it is measured in Columbus / newton.

488. What is 'g' coefficient?

Answer: G-coefficient representing the generated e.m.f gradient per unit pressure input.

489. What is 'h' coefficient?

Answer: 'h' coefficient is obtained by multiplying the g-coefficient by young's modulus valid for the appropriate crystal orientation of the material, and thus measures the e.m.f gradient per unit mechanical deformation, or (V/m) / (m/m).

Multiple Choice Questions

1. The number of hardware interrupts that the processor 8085 consists of is
   a. 1
   b. 3
   c. 5
   d. 7

Answer: c

2. The register that stores all the interrupt requests in it in order to serve them one by one on a priority basis is
   a. Interrupt Request Register
   b. In-Service Register
   c. Priority resolver
   d. Interrupt Mask Register

Answer: a

3. The register that stores the bits required to mask the interrupt inputs is
   a. In-service register
   b. Priority resolver
   c. Interrupt Mask register

d. None of the above

Answer: c

4. The interrupt control logic
   a. manages interrupts
   b. manages interrupt aca5. In a cascaded mode, the number of vectored interrupts provided by 8259A is
   a. 4
   b. 8
   c. 16
   d. 64

Answer: d

6. When the PS(active low)/EN(active low) pin of 8259A used in buffered mode, then it can be used as a
   a. input to designate chip is master or slave
   b. buffer enable
   c. buffer disable
   d. none of the above

Answer: b

7. Once the ICW1 is loaded, then the initialization procedure involves
   a. edge sense circuit is reset
   b. IMR is cleared
   c. slave mode address is set to 7
   d. all of the above

Answer: d

8. Which of the following is not a mode of data transmission?
   a. simplex
   b. half duplex
   c. duplex
   d. semi duplex

Answer: d

Multiple Choice Questions   145

9. If the data is transmitted only in one direction over a single communication channel, then it is of

a. simplex mode
b. duplex mode
c. semi duplex mode
d. half duplex mode

Answer: a

10. The number of hardware interrupts that the processor 8085 consists of is

a. 1
b. 3
c. 5
d. 7

Answer: c

11. The register that stores all the IR in order to serve them one by one on a priority basis is

a. Interrupt Request Register
b. In-Service Register
c. Priority resolver
d. Interrupt Mask Register

Answer: a

12. The register that stores the bits required to mask is

a. In-service register
b. Priority resolver
c. Interrupt Mask register
d. None

Answer: c

13. The interrupt control logic

a. manages interrupts
b. manages interrupt acknowledge signals
c. accepts interrupt acknowledge signal

d. all of the above

Answer: c

14. In a cascaded mode, the number of vectored interrupts provided by 8259A is

a. 4
b. 8
c. 16
d. 64

Answer: d

15. When the PS(active low)/EN(active low) pin of 8259A used in buffered mode, then it can be used as a

a. input to designate chip is master or slave
b. buffer enable
c. buffer disable
d. none of the above

Answer: b

16. The number of hardware interrupts that the processor 8085 consists of is

a. 1
b. 3
c. 5
d. 7

Answer: c

17. The register that stores all the interrupt requests in it in order to serve them one by one on a priority basis is

a. Interrupt Request Register
b. In-Service Register
c. Priority resolver
d. Interrupt Mask Register

Answer: a

18. In a cascaded mode, the number of vectored interrupts provided by 8259A is

a. 4
b. 8
c. 16
d. 64

Answer: d

19. In direct memory access mode, the data transfer takes place

a. directly
b. indirectly
c. directly and indirectly
d. none of the above

Answer: a

20. In 8257 (DMA), each of the four channels has

a. a pair of two 8-bit registers
b. a pair of two 16-bit registers
c. one 16-bit register
d. one 8-bit register

Answer: b

21. For all the four channels of 8257 the common register is/are

a. DMA address register
b. None of the mentioned
c. Terminal count register
d. Status and Mode set register

Answer: d

22. In 8257 register format, the selected channel is disabled after the terminal count condition is reached when

a. Auto load is set
b. Auto load is reset
c. TC STOP bit is reset

d. TC STOP bit is set

Answer: d

23. A circuit that converts n inputs to 2^n outputs is called

a. encoder
b. decoder
c. comparator
d. carry look ahead

Answer: b

24. Encoders are made by three

a. AND gate
b. OR gate
c. NAND gate
d. XOR gate

Answer: b

25. Decoder is a

a. combinational circuit
b. sequential circuit
c. complex circuit
d. gate

Answer: a

26. IC decoders are made with

a. AND gate
b. OR gate
c. NAND gate
d. XOR gate

Answer: c

27. Discrete quantities of information are represented in digital system with

a. Uni code
b. ASCII code

c. Binary Code
d. Octal code

Answer: b

28. Device which converts an input device state into a binary representation of ones or zeros is termed as

a. encoder
b. decoder
c. multiplexer
d. data selector

Answer: a

29. A circuit that changes a code into a set of signals is called

a. encoder
b. decoder
c. multiplexer
d. data selector

Answer: b

30. Modulo 6 counters can be built using a three-element

a. shift register
b. bus
c. flip flop
d. trigger

Answer: a

31. What is a multiplexer?

a. It is a type of decoder which decodes several inputs and gives one output
b. A multiplexer is a device which converts many signals into one
c. It takes one input and results into many output
d. It is a type of encoder which decodes several inputs and gives one output

Answer: b

32. Which combinational circuit is renowned for selecting a single input from multiple inputs & directing the binary information to output line?

a. Data Selector
b. Data distributor
c. Both data selector and data distributor
d. DeMultiplexer

Answer: a

33. It is possible for an enable or strobe input to undergo an expansion of two or more MUX ICs to the digital multiplexer with the proficiency of large number of _____

a. Inputs
b. Outputs
c. Selection lines
d. Enable lines

Answer: a

34. Which is the major functioning responsibility of the multiplexing combinational circuit?

a. Decoding the binary information
b. Generation of all min-terms in an output function with OR-gate
c. Generation of selected path between multiple sources and a single destination
d. Encoding of binary information

Answer: c

35. What is the function of an enable input on a multiplexer chip?

a. To apply Vcc
b. To connect ground
c. To active the entire chip
d. To active one half of the chip

Answer: c

36. One multiplexer can take the place of _____
a. Several SSI logic gates
b. Combinational logic circuits
c. Several Ex-NOR gates
d. Several SSI logic gates or combinational logic circuits

Answer: d

37. Multiplexers work with _____
a. Analog signal
b. Digital signal
c. Both analog and digital signal
d. None of the above

Answer: c

38. TDM stands for _____
a. Time direct measurement
b. Time division multiplexing
c. Time direct multiplexing
d. Time division measurement

Answer: b

39. Which of the following is analogous to multiplexer?
a. Data selector
b. Data multiplexer
c. Data filter
d. None of the above

Answer: a

40. In digital multiplexer selector line is _____
a. Analog value
b. Digital value
c. Unpredictable
d. None of the above

Answer: b

41. Latches constructed with NOR and NAND gates tend to remain in the latched condition due to which configuration feature?
a. Low input voltages
b. Synchronous operation
c. Gate impedance
d. Cross coupling

Answer: d

42. One example of the use of an S-R flip-flop is as _____
a. Transition pulse generator
b. Racer
c. Switch debouncer
d. Astable oscillator

Answer: c

43. The truth table for an S-R flip-flop has how many VALID entries?
a. 1
b. 2
c. 3
d. 4

Answer: c

44. When both inputs of a J-K flip-flop cycle, the output will _____
a. Be invalid
b. Change
c. Not change
d. Toggle

Answer: c

45. Which of the following is correct for a gated D-type flip-flop?
a. The Q output is either SET or RESET as soon as the D input goes HIGH or LOW
b. The output complement follows the input when enabled

c. Only one of the inputs can be HIGH at a time

d. The output toggles if one of the inputs is held HIGH

Answer: a

46. A basic S-R flip-flop can be constructed by cross-coupling of which basic logic gates?

a. AND or OR gates

b. XOR or XNOR gates

c. NOR or NAND gates

d. AND or NOR gates

Answer: c

47. The logic circuits whose outputs at any instant of time depends only on the present input but also on the past outputs are called

a. Combinational circuits

b. Sequential circuits

c. Latches

d. Flip-flops

Answer: b

48. In a J-K flip-flop, if J=K the resulting flip-flop is referred to as _____

a. D flip-flop

b. S-R flip-flop

c. T flip-flop

d. S-K flip-flop

Answer: c

49. How many stable states a combinational circuit have?

a. 3

b. 4

c. 2

d. 5

Answer: c

50. Both the J-K & the T flip-flop are derived from the basic _____

a. S-R flip-flop
b. S-R latch
c. D latch
d. D flip-flop

Answer: a

51. A register is defined as _____

a. The group of latches for storing one bit of information
b. The group of latches for storing n-bit of information
c. The group of flip-flops suitable for storing one bit of information
d. The group of flip-flops suitable for storing binary information

Answer: d

52. The register is a type of _____

a. Sequential circuit
b. Combinational circuit
c. CPU
d. Latches

Answer: a

53. The main difference between a register and a counter is _____

a. A register has no specific sequence of states
b. A counter has no specific sequence of states
c. A register has capability to store one bit of information, but counter has n-bit
d. A register counts data

Answer: a

54. In D register, 'D' stands for _____

a. Delay
b. Decrement

c. Data
d. Decay

Answer: c

55. Registers capable of shifting in one direction is _____

a. Universal shift register
b. Unidirectional shift register
c. Unipolar shift register
d. Unique shift register

Answer: b

56. A register that is used to store binary information is called _____

a. Data register
b. Binary register
c. Shift register
d. D – Register

Answer: b

57. A shift register is defined as _____

a. The register capable of shifting information to another register
b. The register capable of shifting information either to the right or to the left
c. The register capable of shifting information to the right only
d. The register capable of shifting information to the left only

Answer: b

58. In serial shifting method, data shifting occurs _____

a. One bit at a time
b. simultaneously
c. Two bit at a time
d. Four bit at a time

Answer: a

59. A shift register that will accept a parallel input or a bidirectional serial load and internal shift features is called as?

a. Tristate
b. End around
c. Universal
d. Conversion

Answer: c

60. How can parallel data be taken out of a shift register simultaneously?

a. Use the Q output of the first FF
b. Use the Q output of the last FF
c. Tie all of the Q outputs together
d. Use the Q output of each FF

Answer: d

61. The IOR (active low) input line acts as output in

a. slave mode
b. master mode
c. master and slave mode
d. none of the above

Answer: b

62. The IOW (active low) in its slave mode loads the contents of a data bus to

a. 8-bit mode register
b. upper/lower byte of 16-bit DMA address register
c. terminal count register
d. all of the above

Answer: d

63. The pin that disables all the DMA channels by clearing the mode registers is

a. MARK
b. CLEAR

c. RESET
d. READY

Answer: c

64. The pin that requests the access of the system bus is

a. HLDA
b. HRQ
c. ADSTB
d. None of the above

Answer: b

65. The pin that is used to write data to the addressed memory location, during DMA write operation is

a. MEMR (active low)
b. AEN
c. MEMW (active low)
d. IOW (active low)

Answer: c

66. Which of the following is not a mode of data transmission?

a. simplex
b. duplex
c. semi duplex
d. half duplex

Answer: c

67. If the data is transmitted only in one direction over a single communication channel, then it is of

a. simplex mode
b. duplex mode
c. semi duplex mode
d. half duplex mode

Answer: a

68. If the data transmission takes place in either direction, but at a time data may be transmitted only in one direction then, it is of
a. simplex mode
b. duplex mode
c. semi duplex mode
d. half duplex mode

Answer: d

69. In 8251A, the pin that controls the rate at which the character is to be transmitted is
a. TXC(active low)
b. TXC(active high)
c. TXD(active low)
d. RXC(active low)

Answer: a

70. The signal that may be used either to interrupt the CPU or polled by the CPU is
a. TXRDY(Transmitter ready)
b. RXRDY(Receiver ready output)
c. DSR(active low)
d. DTR(active low)

Answer: b

71. The disadvantage of RS-232C is
a. limited speed of communication
b. high-voltage level signaling
c. big-size communication adapters
d. all of the above

Answer: d

72. The USB supports the signaling rate of
a. full-speed USB 1.0 at rate of 12 Mbps
b. high-speed USB 2.0 at rate of 480 Mbps
c. super-speed USB 3.0 at rate of 596 Mbps

d. all of the above

Answer: d

73. The bit packet that commands the device either to receive data or transmit data in transmission of USB asynchronous communication is

a. Handshake packet
b. Token packet
c. PRE packet
d. Data packet

Answer: b

74. High speed USB devices neglect

a. Handshake packet
b. Token packet
c. PRE packet
d. Data packet

Answer: c

75. What will happen if DC shunt motor is connected across AC supply?

a. Will run at normal speed
b. Will not run
c. Will Run at lower speed
d. Burn due to heat produced in the field winding

Answer: d

76. A variable reluctance stepper motor is constructed of _____ material with salient poles.

a. Paramagnetic
b. Ferromagnetic
c. Diamagnetic
d. Non-magnetic

Answer: b

77. DS12887 is a _____

a. Timer IC
b. Serial communication IC
c. RTC IC
d. Motor

Answer: c

78. In a three-stack 12/8-pole VR motor, the rotor pole pitch is

a. 15°
b. 30°
c. 45°
d. 60°

Answer: c

79. What will happen if the back emf of a DC motor vanishes suddenly?

a. The motor will stop
b. The motor will continue to run
c. The armature may burn
d. The motor will run noisy

Answer: c

80. A stepper motor having a resolution of 300 steps/rev and running at 2400 rpm has a pulse rate of- pps.

a. 4000
b. 8000
c. 6000
d. 10,000

Answer: c

81. What is the RAM space that a DS12887 have?

a. 128 bytes
b. from 00-7FH
c. 128 bytes from 00-7FH

d. none of the above

Answer: d

82. If a hybrid stepper motor has a rotor pitch of 36° and a step angle of 9°, the number of its phases must be

a. 4
b. 2
c. 3
d. 6

Answer: a

83. What will happen, with the increase in speed of a DC motor?

a. Back emf increase but line current falls
b. Back emf falls and line current increase
c. Both back emf as well as line current increase
d. Both back emf as well as line current fall

Answer: a

84. In DC motor, which of the following part can sustain the maximum temperature rise?

a. Field winding
b. Commutator
c. Slip rings
d. Armature winding

Answer: a

85. A stepping motor is a _____ device.

a. Mechanical
b. Electrical
c. Analogue
d. Incremental

Answer: d

86. Why two pins for ground are available in ADC0804?

a. for controlling the ADCON0 and ADCON1 register of the controller
b. for controlling the analog and the digital pins of the controller
c. for both parts of the chip respectively
d. none of the above

Answer: b

87. When is the function of the WR pin?

a. its active high input used to inform ADC0804, about the end of conversion
b. its active low input used to inform ADC0804, about the end of conversion
c. its active low input used to inform ADC0804, about the start of conversion
d. its active high input used to inform ADC0804, about the start of conversion

Answer: c

88. State which of the following statements are false?

a. CLK IN pin is used to tell about the conversion time
b. INTR pin tells about the end of the conversion
c. ADC0804 IC is an 8- bit parallel ADC in the family of the ADC0800 series
d. None of the above

Answer: d

89. While programming the ADC0804 IC what steps are followed?

a. select the analog channel, start the conversion, monitor the conversion, display the digital results
b. select the analog channel, activate the ALE signal (L to H pulse), start the conversion, monitor the conversion, read the digital results
c. select the analog channel, activate the ALE signal (H to L pulse), start the conversion, monitor the conversion, read the digital results

d. select the channel, start the conversion, end the conversion

Answer: b

90. Step size is selected by which two bits?

a. Vref/2
b. Vin
c. Vref/2 &Vin
d. None of the above

Answer: a

91. What is the difference between ADC0804 and MAX1112?

a. ADC0804 has 8 bits and MAX1112 has 1 bit for data output
b. ADC0804 is used for adc and dac conversions whereas MAX1112 is used for serial data transmissions
c. ADC0804 has 32 bits and MAX1112 has 3-bit for data output
d. None of the above

Answer: a

92. Which of the following statements are true about DAC0808?

a. parallel digital data to analog data conversion
b. it has current as an output
c. all of the above
d. none of the above

Answer: a

93. Eight input DAC has _____

a. 8 discrete voltage levels
b. 64 discrete voltage levels
c. 124 discrete voltage levels
d. 256 discrete voltage levels

Answer: d

94. INTR, WR signal is an input/output signal pin?

a. both are output
b. both are input

c. one is input and the other is output

d. none of the above

Answer: c

95. The semiconductor memories are organized as _____ dimension(s) of array of memory locations.

a. one dimensional

b. two dimensional

c. three dimensional

d. none of the above

Answer: b

96. If a location is selected, then all the bits in it are accessible using a group of conductors called

a. Address bus

b. Control bus

c. Data bus

d. Either address bus or data bus

Answer: c

97. To address a memory location out of N memory locations, the number of address lines required is

a. log N (to the base 2)

b. log N (to the base 10)

c. log N (to the base e)

d. log (2N) (to the base e)

Answer: a

98. If the microprocessor has 10 address lines, then the number of memory locations it is able to address is

a. 512

b. 1024

c. 2048

d. none of the above

Answer: b

99. In static memory, the upper 8-bit bank of an available 16-bit memory chip is called

a. upper address memory bank
b. even address memory bank
c. static upper memory
d. odd address memory bank

Answer: d

100. In static memory, the lower 8-bit bank of an available 16-bit memory chip is called

a. lower address memory bank
b. even address memory bank
c. static lower memory bank
d. odd address memory bank

Answer: b

101. In most of the cases, the method used for decoding that may be used to minimize the required hardware is

a. absolute decoding
b. non-linear decoding
c. linear decoding
d. none of the above

Answer: c

102. To obtain 16-bit data bus width, the two 4K*8 chips of RAM and ROM are arranged in

a. parallel
b. serial
c. both serial and parallel
d. neither serial nor parallel

Answer: a

103. If (address line) Ao=0 then, the status of address and memory are

a. address is even and memory is in ROM
b. address is odd and memory is in ROM

c. address is even and memory is in RAM

d. address is odd and memory is in RAM

Answer: c

104. If at a time Ao and BHE (active low) both are zero then, the chip(s) selected will be

a. RAM

b. ROM

c. RAM and ROM

d. ONLY RAM

Answer: c

105. A thermistor is a _____

a. sensor

b. adc

c. transducer

d. micro controller

Answer: c

106. What is the difference between LM 34 and LM 35 sensors?

a. one is a sensor and the other is a transducer

b. one's output voltage corresponds to the Fahrenheit temperature and the other corresponds to the Celsius temperature

c. one is of low precision and the other is of higher precision

d. one requires external calibration and the other doesn't require it

Answer: b

107. What is the difference between LM 34 and LM 35 sensors?

a. one is a sensor and the other is a transducer

b. one's output voltage corresponds to the Fahrenheit temperature and the other corresponds to the Celsius temperature

c. one is of low precision and the other is of higher precision

d. one requires external calibration and the other doesn't require it

Answer: d

108. What is signal conditioning?

a. to analyse any signal

b. conversion or modification is referred to as conditioning

c. conversion from analog to digital is signal conditioning

d. conversion from digital to analog is signal conditioning

Answer: b

109. What steps have to be followed for interfacing a sensor to a microcontroller 8051?

a. Make the appropriate connections with the controller, ADC conversion, analyze the results

b. Make connections of 8051 with an ADC to change analog to digital, send this value to the controller, analyze the results

c. Interface sensor with the MAX232, send now to microcontroller, analyze the results

d. None of the above

Answer: b

110. LM35 has how many pins?

a. 2

b. 1

c. 3

d. 4

Answer: c

111. Why reference voltage of ADC0848 to 2.56 V if I/P is connected to the temperature sensor LM35?

a. to set the step size of the sampled input

b. to set the ground for the chip

c. to provide supply to the chip

d. all of the above

Answer: a

112. How many clock pulses are confined by each machine cycle of Peripheral-Interface Controllers?

a. a.4
b. 8
c. 12
d. 16

Answer: a

113. Which flags are more likely to get affected in status registers by Arithmetic and Logical Unit (ALU) of PIC 16 CXX on the basis of instructions execution?

a. Carry (C) Flags
b. Zero (Z) Flags
c. Digit Carry (DC) Flags
d. All of the above

Answer: d

114. What is the execution speed of instructions in PIC especially while operating at the maximum value of clock rate?

a. 0.1 μs
b. 0.2 μs
c. 0.4 μs
d. 0.8 μs

Answer: b

115. Which operational feature of PIC allows it to reset especially when the power supply drops the voltage below 4V?

a. Built-in Power-on-reset
b. Brown-out reset
c. Both a & b
d. None of the above

Answer: b

116. Which among the below stated reasons is/are responsible for the selection of PIC implementation/design on the basis of Harvard architecture instead of Von-Newman architecture?

a. Improvement in bandwidth
b. Instruction fetching becomes possible over a single instruction cycle
c. Independent bus access provision to data memory even while accessing the program memory
d. All of the above

Answer: d

117. Which among the below specified major functionalities is/are associated with the programmable timers of PIC?

A. Excogitation of Inputs
B. Handling of Outputs
C. Interpretation of internal timing for program execution
D. Provision of OTP for large and small production runs

a. Only C
b. C & D
c. A, B & D
d. A, B & C

Answer: d

118. What is the status of shift clock supply in an USART synchronous mode?

a. Master - internally, Slave - externally
b. Master - externally, Slave - internally
c. Master & Slave (both) - internally
d. Master & Slave (both)- externally

Answer: a

119. Which bit plays a salient role in defining the master or slave mode in TXSTA register especially in synchronous mode?

a. RSRC
b. CSRC
c. SPEN
d. SYNC

Answer: b

120. Which register/s should set the SPEN bit in order to configure RC7/RX/DT pins as DT (data lines)?

a. TXSTA

b. RCSTA

c. Both a & b

d. None of the above

Answer: b

122. Which instruction is applicable to set any bit while performing bitwise operation settings?

a. BCF

b. BSF

c. Both a & b

d. none of the above

Answer: b

123. Where is the result stored after an execution of increment and decrement operations over the special - purpose registers in PIC?

a. File Register

b. Working Register

c. Both a & b

d. none of the above

Answer: c

124. Which flags of status register are most likely to get affected by the single cycle increment and decrement instructions?

a. P Flags

b. C Flags

c. OV Flags

d. Z Flags

Answer: d